PENGUIN BOOKS

THE VISION OF GLORY

'John Stewart Collis is a poet, a scientist, a scholar. His divine gift is to explain the *extraordinary nature of the ordinary*. He has faithfully and gladly pursued a quest which, starting with physics, ends quite naturally with metaphysics. Thrilling beyond any thriller' – Maurice Wiggin in the *Sunday Times*

'He is the poet among modern ecologists, a natural philosopher who, whether he is writing about trees or rainbows, an iceberg or a piece of chalk, never takes a fact without linking it to an idea, or an idea without connecting it to a fact. His book dispenses information in the language of the imagination, and by peeling back the film by which everything appears dully familiar, reveals a vision of the world miraculously transfigured' – Michael Holroyd in *The Times*

John Stewart Collis was born in 1900 of an Irish family. He was educated at Rugby School and Balliol College, Oxford. Among his publications are *Shaw* (1925), *The Sounding Cataract* (1936), *Down to Earth* (1947), which won the Heinemann Foundation Award, and *The Life of Tolstoy* (1969). His latest book is *The Worm Forgives the Plough* (1973), which is also available in Penguins.

JOHN STEWART COLLIS

THE VISION OF GLORY

THE EXTRAORDINARY NATURE
OF THE ORDINARY

WITH A PREFACE BY
RICHARD CHURCH

PENGUIN BOOKS

Penguin Books Ltd, Harmondsworth, Middlesex, England
Penguin Books Australia Ltd, Ringwood, Victoria, Australia
Penguin Books Canada Ltd, 41 Steelcase Road, West, Markham,
Ontario, Canada
Penguin Books (N.Z.) Ltd, 182–190 Wairau Road, Auckland 10,
New Zealand

—

First published by Charles Knight, 1972
Published in Penguin Books 1975

—

Copyright © John Stewart Collis, 1972

—

Made and printed in Great Britain by
Cox & Wyman Ltd, London, Reading and Fakenham
Set in Linotype Times

CONTENTS

CONTENTS

BOOK II: THE MOVING WATERS

CONTENTS

CONTENTS

PREFACE

by Richard Church

I should call John Stewart Collis a 'natural philosopher' in the old-fashioned sense of that phrase. There were natural philosophers in the eighteenth century. Then, in the nineteenth, the idea of one man being able to comprehend the whole was frowned upon. Specialization set in: a dangerous mode which today threatens us heavily because it has led to indiscriminate superstitions and hierarchies. Outstanding as a rebel against this tendency is John Steward Collis. I suspect that Time will select his trilogy *Paths of Light*, *The Moving Waters*, and *The Triumph of the Tree* as a lasting achievement.

Paths of Light is his most ambitious atempt to connect knowledge. The subject inevitably embraces the new physics, and his chapter on this is one of the most exciting things he has ever done. The book demonstrates what I would call 'The Dance of Life', the miracle by which the animate design of the universe is maintained, from the vast orbits of the galaxies to the equally lonely minuscular movements within the atom of hydrogen.

In *The Moving Waters*, where he assumed a stance comparable to that of Thales, the philosopher of early Greece, he revealed the function and garments of water in the scheme of things. In a succession of dramatic pictures he moves from the primordial through the manifestations down to the tiny drop of dew. Here is one picture:

If vapour were not as invisible as wind (except when it chooses to show itself), and could we stand on a high place commanding an immense view of land and sea, we would behold on a sunny day fountains rising into the sky from ocean and lake, from river and pond, from forests and woods, from single trees and flowers, from copses and fields. When we are looking at clouds we are not looking at things which belong to the sky; we are looking at the waters of the land and of the sea. They have abandoned their stations. They have strayed from their courses. They have passed from their confinement in the restricted river and the level lake and the restless ocean, they have left their corridors in the fluted trunk and the waving grass and the phalanxed corn, and have risen into the firmament to assume another aspect . . .

It is a prose that may remind the reader of Ruskin, and make him realize that here is a poet at work, who has been reverent before the display of contemporary knowledge in the sciences, trying to correlate their rich offerings to interpret afresh from that map the ever-increasing intricacies and wonders of daily life.

The Triumph of the Tree set out to show what part vegetation plays in the world, and how it affects human life. It gave mankind some shrewd knocks about the abuse of vegetation, and what damage we have done, and continue to do, to the face of the earth by our ignorant methods of cultivation and our selfish indifference to the needs of posterity. Collis has contrived, by some remarkable touch of poetic skill, to make us *see* the forests of the world. It is a wonderful experience to read and to become aware in the reading that the plant life over the face of the earth is sentient, partaker in a scheme of growth and evolution whose subtlety of detail epitomizes at any given moment the ultimate pattern. That visual genius (it is no less) animates each of these books. And it has to be emphasized that throughout Collis is dealing with the reality *behind* the tangible. He is seeking out the *meaning* as well as the *pattern* to the unfolding buttercup petals, or the encroachment of a hundred square miles of forest across what was formerly moorland. And he makes us watch these movements by the slow-motion pictures recorded in the camera of his extraordinary imagination. It is a religious imagination, concerned always to link things together, to make them explain each other and each other's conduct in this weltering dance of physical life.

BOOK I
PATHS OF LIGHT

Facing the problem of space I have sought for sequence as my chief aim. Having entirely excluded anything from *While Following the Plough* and *Down to Earth* (though they may be my best efforts) I am free to deploy a good deal in the room allotted to me. From *Paths of Light* I have taken the whole of the first part – for I must start this book with the Atom – and, while omitting 'The Story of the Lamp', the whole of the last part – only leaving out the middle chapters. This assures considerable sequence.

– JOHN STEWART COLLIS

PART ONE
LIGHT

CHAPTER I
THE NEW ALCHEMY

1. The Modern Discovery

We look out upon the world and we see the firm ground beneath our feet, the weighty mountains, the waters, the skies, the plants, the animals, and the human race. There is a difference, we notice, between the plants and any piece of solid ground – the plants get up and they fall down. They possess a force of some sort: whereas we never see a rock swell and shrink in this manner. We notice that animals have the power not only to grow but to move. It is the same with ourselves: we can lift our arms: we can go about – in answer to a principle of energy. We observe another force at work – the strong strokes of the wind. We see another – the power of water as it flows or falls. And another – fire leaping from object to object. And again – a tree struck down by a flash of lightning.

So much has been obvious to man since before the beginning of history: on the one hand the solid, inanimate earth; and on the other, Force disclosing itself as motion in the growing plant and the moving creature, in wind, water, electricity, and the consuming fire.

From the dawn of history men were aware of these powers, and as time went on they began to add the force of wind and water to the energy of their arms. Then they went a step farther and used fire, and farther still and used the swifter combustion of gases, and at the same time were enabled to channel the currents of electricity. There they stopped.

They stopped there until only the other day. If we pass in

review all the civilizations which have risen and fallen since the first records, we see that not one of them, and not our own civilization until about half a century ago, knew that more energy was to be found elsewhere. After all, why should there be? The universe was getting on all right. It didn't need any more force, and in any case there was nowhere to look for it.

This was an error, as everyone now knows. There is force in reserve. There is force deposited in a bank, as it were. That bank is matter. It is now known that matter is not inanimate, that it is not inert, that it is not lifeless. The material composing the mountain, the lake, the air, is made up of cages imprisoning lions and tigers.

This is the great modern discovery – with which nothing can compete. It was not astonishing to find that the moon is rather more than twenty miles away, and that the stars are larger than sixpenny pieces. It was not astonishing to learn the facts about combustion, magnetism, electricity, or gravitation, since they openly declared themselves; but it is astonishing to discover that this stone which I hold in my hand is not a solid but a conglomeration of cages, each with a tiger inside. I put it that way because I see it that way, and I need strong words, though those words are too weak and none could be obedient enough to do justice to the full truth, and I am amazed at the indifferent manner in which people today accept as a matter of course the discoveries of nuclear physics. It is as if there were a determination in modern times *not* to regard the mysteries of the world as mysteries, and to discourage any movement of the imagination. It is allowed that a 'heavenly vision' may come to mystics and seers from the view on the mountain-top, but a like vision is not supposed to follow from a view of the less easily discernible lands of the nuclei where nevertheless the builder has laid the foundations of empire. Physics is not supposed to be a religious or a poetic study. I cannot share this view, and I confess that ever since I looked into the subject hardly a day has passed as I have gone about that I have not increasingly marvelled at the masonry of creation and the mystery of design. Dr George Gamow, celebrated physicist and delightful writer, when writing his book on the sun, found himself obliged to devote some space to physics. He apologizes for this. 'The author regrets the pain that this excursion into the domain of pure physics may cause some readers who picked up this book for its astronomical title,' he says, 'but except for poets, no one should speak about stars without knowing the properties of matter of which they are constructed.'[1] I am sorry he said that about poets. If such knowledge is to be considered too stiff for those who make the poetic approach to reality, then the absurd position may yet be reached when science can give nothing to poetry and poetry nothing to science.

2. The Atom

Let us take the nearest thing that comes to hand – a stone. Until roughly a hundred years ago in spite of the penetrating deduction of Democritus among the non-experimenting Greeks, it had been generally believed that such a piece of matter was a continuous whole, as you might think of a slab of jelly; and that theoretically you could divide it up into ever smaller and smaller and smaller pieces. It was not till the middle of the nineteenth century that the discovery was made (by Dalton) that matter is already divided up, and is not like a slab of jelly but a packet of peas. These peas were regarded as final, and called atoms, after the Greek – 'that which cannot be divided'.

The scientists then set to work to find out what they could about these atoms, and soon ascertained that they were not all the same size nor all the same make. Their size differed no more than apples on an apple tree, but some of them differed from others so much that if a sufficient number of a similar lot were bunched together to become visible, you saw gold, while another lot would give you mercury. It was found that there were ninety-two different sorts (a few more were added later) and these were now named the Elements – thus disposing of the old time-honoured idea that there were just four fundamental elements, Fire, Earth, Air, and Water. The reason why we see

1. *The Birth and Death of the Sun.*

so many more substances in the world than ninety-two is because the different sorts of atoms combine as molecules to make a great variety of compounds, the most famous example being when two atoms of hydrogen join with one atom of oxygen to give us something quite different from either, which we call water: while a more complicated example of atomic building would be, say, Socrates.

Their minuteness offers a pictorial challenge. 'No such things as atoms?' said Rutherford. 'Why, I can see the little beggars!' But that is just our difficulty; we cannot see them, nor can any ordinary microscope, and we are obliged to be content with utterances of those who can make formidable mathematical calculations. Thus we are told that the number of atoms necessary to cover the space on a speck of dust would be some thousand million million – 1,000,000,000,000,000 atoms perched on a speck of dust. Take two more examples. Remembering that a molecule of water, equals three atoms, 'The number of molecules in a little drop of water', says Gamow, 'is about the same as the number of drops of water in the great Lake Michigan.' And this from the accomplished writer as well as physicist, Professor E. N. da C. Andrade: 'If a staff of a thousand men were told off to count the atoms in a single one of the little bubbles of gas which collect on the side of a glass of soda-water, and if each man could count three hundred atoms a minute, and counted twelve hours a day all the year round, the job would take a million years.' That seems strange to me. It is almost saying that there is no limit to smallness. I find it easier to imagine no limit to bigness. There must be a point when something becomes nothing, and one cannot help feeling that some things would have become no things before a thousand million million of them could find room on a speck of dust. Is it a myth? Is it a fairy tale to end all fairy tales? Are we to trust our scientists more than the Schoolmen of old who debated on how many angels could balance on the point of a needle? We are inclined to ask such questions in our surprise; yet we are bound to acknowledge that we are scarcely entitled to use the word myth in this connection. A myth is something which is not objectively true. The fairies were not objectively true, the host of primitive gods were not objectively true, they were mythical, and in consequence though you prayed to them and performed rituals in their honour and made sacrifices to them, you could never count upon their doing what you asked, since unfortunately they did not exist. Our scientists are on firmer ground. They say there is a genie in the bottle. If we say – Nonsense, there is no bottle and no genie, they can uncork it and let out a monster who in a few minutes can destroy a city. But this is anticipating – we are bound to accept atoms and a lot of them, so perhaps it does not matter if we accept a greater number than we can conceive.

We do see atoms whenever they are bunched together in sufficient numbers to show us a substance; but as we cannot see them individually we find it hard to realize that they are all *in motion*.[2] They are all dedicated to eternal activity: unpausingly they pursue an endless path and revolve in ceaseless chase – in the gas, in the liquid, in the solid. It is as if matter were composed of untold numbers of swarming insects. We can accept this as we walk through a gas, since they are then free from one another; we can accept it as we swim through water, since they are then fairly free; but how about the stone wall – surely there they are so clamped together that they have come to rest? Yet no, even there they are pushing and pulling, quivering, squirming, vibrating, struggling – for solids are by no means solid. We should really expect this activity on the part of atoms, for motion is Energy made manifest, and atoms are the immediate ministers of Energy. Yes, we may say, we can accept that, but accustomed as we are to think of nature in terms of mighty opposites such as light with darkness, heat with cold, noise with silence and so on, should we not find the opposite to motion, which would be perfect stillness? Yet we are told that it is not in the solid, not in the corpse, not in the tomb. Is there then no standstill? There is: but we can only find it in a temperature of 273 degrees centigrade below the freezing point of ice – when the very air itself would be a concrete block.

2. *Known as the Brownian Movement.*

3. The Atom within the Atom

For a decade the scientific world was content, in general, to accept atoms as final in themselves – as indivisible particles, solid as billiard balls. I say in general, for some, especially Clerk Maxwell, were not prepared to accept it save under protest. For if the doctrine of evolution were true and Nature proceeded by process of steady building and change, how could the foundation bricks consist of lifeless, changeless particles? And how on earth could they cling together and form wholes? Nevertheless it did not appear to trouble people very much, and it seems to have been accepted as quite in order by the main body of scientists and laymen alike.

Then came the great day when one of the elements was discovered to be buzzing with *interior* activity. It was found to be pouring particles out of itself at a terrific speed, rather like a volcano in eruption, or as if we had come upon a rock which turned out to be a fort firing into the open. Such a fort could not be solid, it must contain room for guns, ammunition, and soldiers. This was the element of uranium thus found to be so active – or radioactive as we now say. It was immediately deduced that if one element were like this the other elements probably possessed equally interesting interiors. At once Rutherford said, in effect: 'We must penetrate the other atoms. How can we do this? We cannot unscrew them with tweezers, we cannot split them open with a knife – but we can take a gun and shoot at them. And what better gun can we use than a radioactive substance, and what better bullets than the particles shooting out at almost the speed of light?' A good idea. We can all have good ideas. 'Do you think highly of *Gulliver's Travels*, Dr Johnson?' someone asked. 'Why, sir,' he replied, 'when once you have thought of big men and little men, it is very easy to do all the rest.' Yet of course the truth is the opposite to this. We can all think of big men and little men, and all fail to do anything with them. That is the history of failed artists and scientists and inventors the world over. It is not the history of

Rutherford, whose largeness and geniality loom even in every textbook which mentions his name. He took the gun and fired with the alpha particles (as they were called to distinguish them from two other rays that also came out) at atoms in his laboratory at Cambridge. Nothing happened. He fired again. Nothing happened. He fired again, a machine-gun fire of ten thousand bullets. Nothing happened. He had shot at ghosts. He persevered – and suddenly one of the particles hit something and bounced away. From that moment, we may say, the door was open and a thorough-going examination of the interior of the atom became possible.

I will set down the relevant facts: they can be stated quite briefly without doing violence to the complexities which we accept as a matter of course. One point ought to be made clear first. The layman is, I think, inclined to be a little misled about one thing by the scientists, on account of their terminology. They speak of the 'rim' of the atom, and the 'shells' of the atom, and the 'flesh' of the atom, not to mention the 'splitting' of the atom, as if it were a ball or a box with a lid or a covering. But is it like a box that you can open? or an onion that you can peel? or a fleshy oyster with a pale poetic pearl for a prize? By no means. Its boundary consists of whirling electrified particles – which alone are responsible for 'rims' or 'shells'. The following is all that need be said at the moment about the structure of the atom. It contains a central core round which other particles circle at the speed of several thousand million million revolutions a second. They travel in a circle because they are attracted by this central nucleus rather as the earth is attracted by the sun, but not for the same reason. The attraction here is electrical; for the machinery of the universe is largely governed by the fact that one of its fundamental forces, electricity, attracts and repels itself by virtue of two manifestations of itself. One of these manifestations we call positive and the other negative, though the terms are as arbitrary as if they had been called male and female, there being nothing more negative about the negative charge than the positive. In fact both are extremely positive in this – that a positive charge repels a positive charge and a negative a negative, while a positive attracts a negative and a negative a positive, and the attraction is so enormously greater than that of gravitation as to promote general harmony and stability in the architecture of matter. The central core, or nucleus, or atom within the atom, is positively charged with particles called protons while the encircling particles called electrons are negatively charged, and being much lighter would certainly be drawn into the centre were it not for their excessive speed. This speed also serves to clear a space round the nucleus and prevents other atoms from coming too near, for it creates a kind of rim or round wall, and the orbit of this wall defines the size of the atom. In the main this will pass as the model for all atoms from the simplest hydrogen atom with one proton balancing one electron, to uranium with ninety-two of each, though extra neutralized particles called neutrons generally add further weight to the nucleus.

Two more facts of interest arise from this structure. First, practically all the weight of the atom resides within the nucleus It takes millions of nuclear particles to occupy a cubic inch, but that cubic inch would weigh a billion times as much as a cubic inch of water – that is about ten million tons. I take that figure from Hecht. Gamow formulated, with practical success, the theory that different nuclei may be considered as the minute droplets of a universal 'nuclear fluid', and he calculated that one cubic centimetre of it would weigh two hundred and forty million tons. So we are in fairyland again: at least not the land we experience where a pound of sugar is quite a large object.

The second fact of interest is that the volume of space kept clear by the electrons is enormously greater than the total volume of the electrons themselves – the ratio being that of bullets in a battlefield. The nucleus, though very heavy, is very small – less than a single electron. Thus the space between the orbit and the nucleus is so large as to jusify the comparison that, say, the six electrons in the atom of carbon are like six wasps in Waterloo Station – with a fly in the middle representing the nucleus. No wonder Rutherford was obliged to fire ten thousand bullets before he hit his target.

Thus we can see that a solid is not solid but chiefly holes; and water also full of holes; and the air riddled with holes. It would be interesting to be able to make practical demonstrations of this, for if we could empty the emptiness out of an elephant it would shrink to much less than the size of a mouse; indeed, they say that if the Empire State Building in New York were thus treated it would lodge comfortably on the head of a pin – though it would not be easy for us to pick up the pin-head since it would weigh tens of thousands of tons. It is not easy for us to realize the essential hollowness of things, that we are all hollow men. We see a great hulk of material in front of us and we suppose that it is massive because of its material. Yet it is massive because it lacks material; it is bulky because it has little bulk; it is visible because most of it does not exist. This is not immediately obvious. But if we make a pyramid of empty petrol-tins and then squash them together leaving no hollowness, we would see a remarkable shrinkage of the pile. How much more if we squash together the millions and millions of tiny hollows in every thousandth of an inch of material. The truth is that emptiness is the norm of the universe. It is almost void of matter. We think of the constellations of the heavens and of the countless stars in the uncounted galaxies. Yet a cosmic ray travelling at the speed of light for ten thousand million years is unlikely to traverse as much matter as would cover a two-shilling piece. Jeans offers the pictorial analogy that six specks of dust in Waterloo Station represent the extent to which space is populated with stars. If we turn our gaze from the unspeakable amplitude of the sky to the minute particulars of the earth and remember those six wasps in Waterloo Station as representing six electrons in the atom of carbon, we begin to realize that though the universe is big the difficulty is to find anything in it.

4. The Bonds

Yet – has it not been said? – 'Where there is Nothing there is God.' We do not understand the mystery of Nothingness. We cannot grasp the creativity of emptiness. 'Clay is fashioned, and thereby the pot is made; but it is its hollowness that makes it useful,' said the great mystic, Meister Eckhart. 'By cutting out doors and windows the room is formed; it is the space which makes the room's use. So that when things are useful it is that in them which is Nothing which makes them useful.' All we know is that when we contemplate the universal void, the fields of space, the everlasting horizons of unlimited nothingness, certain Manifestations, which we call life and matter, appear before our eyes. Given the delicate labours of the physicists it is the privilege of modern man to behold the bricks which have been appointed for the foundation of the earth. If from the mountain-tops we can sometimes be lifted up in spirit as we survey the finished figures, our elation need not be less when we look down and mark their masonry.

There are the ninety-two elements. We must not think of them, and we do not think of them, as sitting in a row independent of one another and sufficient unto themselves. They combine. Just as the twenty-six letters of the alphabet, which on their own are not much good, are combined to give us words and all the wonder of words, so the ninety-two elements (or most of them) combine to give us the wonders of the world. The question is – why do they combine, and how do they cohere?

How do they manage to stick together? The packet of peas, as it were, should spill all over the place. There is no glue holding them in position. If we shatter a stone we cannot bring the pieces together again. Many answers to this question were advanced at first. Some experts held that the atoms had claws or brackets, others said that they possessed a sort of spear with which they hooked each other, while yet others maintained that they came together because they loved one another, and we were given what might fairly be called the love-life of atoms.

We smile, but I think we must agree that attraction, magnetism, affinity, love, such as we think of it in human terms need not be denied roots reaching back into the abyss of time. Still, to use the word 'love' here would lead us straight into superficial mystification and not in the least into profound mysticism. Let us use the right words in the right places. The right word here is 'electricity' – by no means to be equated with Spirit.

It is electricity which is mainly responsible for bringing the atoms together and for holding them in place; and it is the electrons – ridden as it were by electricity like a jockey on a horse – which are the agents of combination and cohesion. Again let us consider the ninety-two elements. If we make a list of them and arrange it according to weight and electronic number from the simplest to the most complex, we will notice that the most stable elements appear at mathematically regular intervals and possess certain attributes, and for this reason we call our list the Periodic Table. All the elements are not stable. Many are in danger of breaking down if left to themselves, of losing their identity, and the more complicated they are in their rings or shells of electrons, the more liable they are to disintegrate like uranium and radium at the far end of the Table, which cannot preserve their integrity and wither into lumps of heavy hopeless lead. 'The overcrowding in these radio active elements', says Bertrand Russell in an artless sentence, 'must be something awful. No wonder they are in a hurry to leave such a slum.'[3] The element of silver which comes in the middle of the Table is said to occupy the place of greatest stability, and it seems that if there were more instability among the other elements the whole earth would be made of silver. Apart from this, those elements which occur in set periods are so much more stable and independent that they are called Noble or Inert Gases, all of which except one possess the important property of eight electrons in their outer rings. Now there is such virtue in this number eight (I am sorry it is not seven) that all the other elements strive towards the octet so as to achieve stability. They manage this by joining forces and

3. *The ABC of Atoms.*

exchanging electrons or, if need be, sharing them. Thus it is easy to see that lithium with only one electron in its outer shell will readily combine with fluorine, which having seven needs one, and so both are satisfied in their search for stability, for both in combination have achieved the octet; that when an oxygen atom, which needs two electrons, takes them from beryllium which has two to spare, then both are satisfied in their search for stability, for both have achieved the octet; that when an atom of sodium which needs to get rid of one electron in order to become like the Noble Gas neon, exchanges it with chlorine which needs one more in order to become like the Noble Gas argon, so that together they form a molecule of an entirely new property called salt, then both are satisfied in their search for stability, for both have achieved the octet.

I lay down the general principle which we can all understand, without the blur of modification and with no attempt even to suggest the details of the dance of the electrons, of the electrical ballet in the orbits of all shapes and sizes from swelling circle to narrowest ellipse, from solitary path to crowded causeway, for neither the imagination, the eye, nor the microscope can help us to do justice to the reality. The notable thing is that though the electrons move round their orbits about seven thousand million times in the millionth of a second, and we should expect endless collision and hopeless chaos, there is instead the beauty and rest of perfect symmetry.

Thus the elements are drawn together in fraternal embrace in every kind of combination from simple to elaborate forms, and we get the world we know. We get more – for we promote further assemblies ourselves. We create more world. This comprises the achievement of chemistry, revealing the power of man to understand the movements and anticipate the needs of the elements. When we think of this, when we contemplate the work of these architects who build with invisible bricks, we feel that man with all his faults, with all his clay, his mud, and his madness, has truly in him, as Nietzsche said, 'the creator and the sculptor, the hardness of the hammer, the divine blessedness of the spectator on the seventh day'. But my main concern, the main point which I have reached at last, is the fact that this interlocking of the outer electrons of atoms serves to forge very powerful *bonds*.

5. The New Alchemy

It is not easy to break these bonds. If they are cast asunder, if they are snapped, as you might snap a taut piece of elastic, or wound-up spring, then energy will be released, and the amount of that energy will be according to the strength of the bond.

This seldom happens in the ordinary course of nature, for stability has been established in the compounds, as we have seen. It does happen occasionally as when lightning, for instance, causes wood to 'catch fire' as the strange phrase goes. Actually it does not catch anything, it is transformed into fire on account of the extraneous force snapping the bonds and accelerating the motion of the atoms of oxygen to form the new compound of carbon dioxide. This transmutation, this electrical reaction, is known as a chemical reaction, and it is upon such liberations of energy that the whole of modern civilization is built up. In our search for power we have been able to use the slow reaction from the unstable compound of coal, and get heat to change water into steam that will push a piston and drive an engine; or we can release energy in a vaporized combination so quickly in a series of combustions that we can dispense with the intermediary of water and steam; or we can release energy in a mixture of sulphur, charcoal, and saltpetre so very much more quickly that it will burst its container in an explosion. But the compounds of the world are nearly all stable. We cannot easily loose the bonds of the minerals, we cannot burn stones or metals. By a lucky chance we can liberate the energy and thus destroy by transmutation the substance of coal which is squashed primeval forest, which on account of certain geological conditions did not slowly burn (or rot) in to carbon dioxide; and we can liberate the energy in oil which by another lucky chance is squashed marine organisms.

It was clear that the chemical sources of energy were limited, and that we might soon be compelled to return again to the unlimited supplies of wind and water. Then, at the turn of this century, we discovered the inner bonds of the atom. Before the discovery of radioactivity the play of the outer electrons had been recognized sufficiently to promote the application of chemistry; but it was not known that another atom, as it were, existed within the atom itself – the final nucleus. The famous 'splitting' of the atom really means opening the inner sanctuary of the nucleus. The radioactive elements showed that it was sometimes splitting open on its own account and at its own pace, and incidentally causing the transmutation of the element involved. I say 'incidentally', though there was a time when the possible transmutation of elements was a wonderful dream. The alchemists of old clung to the idea that such a thing was possible, and in their primitive laboratories strove to turn base metal into gold. We can do this today: at least theoretically we can, but we do not want to, and might just as well turn gold into base metal, since any element from which we can release nuclear energy is worth many times its weight in gold – for while a gramme of gold is worth about seven shillings, a gramme of lithium is worth about three thousand pounds.

That pound's worth of power can be got out of a gramme of lithium. We have noted how the amount of energy released in a chemical reaction corresponds to the strength of the bonds. But the binding of the outer electrons is nothing compared with the strength of the inner nuclear knots. The question was how to untie them, for it was one thing to break a molecule and get a chemical reaction, but quite another to break a nucleus and get an alchemical reaction. Even Rutherford thought it could not be done. 'It is like trying to shoot a gnat in the Albert Hall in pitch darkness,' he said to Mr Ritchie Calder, 'and using ten million rounds on the off-chance of hitting it.' But, as we know, Sir John Cockcroft was not of that opinion, and other methods have now been devised by means of which the nuclei are encouraged to shoot at each other in a kind of furnace called a cyclotron.

At first we may feel surprised that so much power should reside in this minute place – any power at all, let alone more than that of ten million tigers. We are surprised because we have been slow to learn what energy actually *is*. We must remember that the nucleus is the only part of the earth where matter really exists – all else that we see is façade and fable, all else is pretence, posturing, hollowness, sham. Here only is true concentration, here only the clenched embrace of matter when it can truly be called mass. And what is mass? Einstein came forward and said $E = M$. A great saying: one of those supreme utterances when the revolutions of nations are already decided and histories unwritten are written. Energy equals Mass. Mass is Energy made manifest in concrete form. Energy is not indestructible. Matter is not indestructible. One can be converted into the other like work and heat. We do not need a lot of mass (or what we in our ignorance think of as mass) to get a lot of energy. The loss of mass which will give us an alarming amount of power cannot be discerned by the naked eye. Given the equation and the high mechanical knowledge required; given the equation, more strictly stated as $E = mc^2$ (meaning that the loss of mass must be multiplied by the square of the velocity of light), then it is possible to calculate a fixed scale of convertibility. We can calculate that an alchemical reaction is ten million times greater than a chemical reaction in terms of combustion. We can calculate that one gramme of uranium harbours energy equal to nineteen tons of T.N.T., and that it will give a million horse-power for thirty-three hours – which is what America and Canada jointly draw from Niagara in that time. We can calculate that one-tenth of an ounce of radium will give out the energy of a ton of coal. We can calculate that since one pound of coal burned at Battersea Power Station keeps one electric fire going in one household for four hours, then the same lump of coal if converted into direct energy would keep all the electric fires in all the households in the United Kingdom going day and night for a month. Consider the omnipresence of this power. It lies hidden in the fortress of all the nuclei of all the atoms; it lies sleeping within every thing, within every stone, within every piece of bread and lump of cheese, within ourselves!

Our immediate task is over. Our goal is reached. We open the furthest door and behold the last links of the mighty locks. Here, in this unlikely place, we come upon the throne of majesty; here, concealed from mortal sight, dungeoned from the light of day, wholly lacking in all the props and appointments of power, in these forever exiled halls – the final, awful essence sits in state. It was not surprising that there was some heart-searching about this at first. 'Canst thou bind the sweet influences of Pleiades, or loose the bands of Orion?' asked God of Job when rehearsing the grandeur of His conception and the amplitude of His arm. Should we loose the bonds that hold the earth together, thus unstabilizing the stability? We have wondered about this. And it shall be written of our day, that having unfolded and still unfolded the mystery of matter until he came to stand before the last stronghold, Man paused. He hesitated. Should he take this citadel? Should he enter in at this gate? When at length, in August 1945, it became possible for him to use the knowledge, he said – *No: we should not have come to this place*. And so thinking, he turned and loosed these powers against himself.

6. The Sun

Still, Hiroshima may prove to be the pea-shooter's return. At least there are real signs that mankind is afraid of nuclear war and may now stop. I live on that assumption myself. I am more encouraged to do so since the arrival of Zeta means that the physicists have not only given us a good reason for not waging war, but have taken away the main object of conquest – material gain in terms of gold or coal or oil or slaves. Uranium is a scarce commodity; but if in future we can get all the power we want by dipping a cup into the sea we need never go to war to plunder that. And I am not prepared to argue on rational grounds that we should not interfere with the nucleus, since there is no more *reason* why we should refrain from snapping the inner bonds to give us an alchemical reaction than from the now time-honoured practice of snapping the outer bonds to give us the ordinary chemical reaction.

What immediately concerns us in this study is the fact that atoms are the source of light. 'Every kind of light has its beginnings in atoms.'[4] We think of light as primarily coming from the sun. But sunlight also comes from the earth – or can be made to do so. For the earth is a bit of the sun which has 'gone out'. The sun consists largely of the same materials as the earth in a chain-reaction of transmutation. It is therefore not surprising that we can turn pieces of the earth into sun again. An atomic pile can do this. The sun is a glorified atomic pile.

I must expand this last statement. During some hundreds of years the chief question concerning the sun which exercised the minds of the scientists was: How does the sun keep alight? The earth was a piece of the sun thrown into space. It cooled. Why does the sun not cool? What keeps it going?

As is usual in the history of science many ingenious, imaginative, plausible, and even amusing theories were advanced in answer to this. Thus they said at first that it was simply a gigantic fire getting its energy by combustion like a burning pile of coal. Then they calculated that if this were so, and even if it were burning best-quality Newcastle Coal, it would peter out after five thousand years. (I would have said five months.) They added that if it were using coal as fuel the heat would be only one millionth part of the energy actually emitted. After this Hermann von Helmholtz came forward with his gravitational contraction theory which amounted to the idea that the terriffic pull of gravitation would cause such a falling inwards as to generate energy, and that this would account for sunshine already having lasted for two thousand million years. But since this would mean a shrinkage of the sun's radius amounting to 0.0003 per cent every century, it was realized that such rapidity ruled the theory out of court. This was followed by what may fairly be called the Perpetual Accident idea, which was that the

4. E. N. da C. Andrade: *The Atom and Its Energy.*

sun did receive fuel from outside itself and was in fact being steadily stoked up by the fall of meteors upon its surface. In order to maintain a balance between income and expenditure the accretion would have to be about equal to the loss in radiation. In view of the fact that such addition should be roughly four million tons every second, the theory was soon dropped.

Happily, speculation is now at an end and there is agreement as to the solution of this problem. It is now understood that the sun is not on fire: that the sun is not burning. It is hot; but it is too hot to burn. We cannot strictly use the word burn in this connection, nor speak of chemical combustion. Its energy is recruited by the transformation of hydrogen into helium. There was a time when seeing a picture of some huge atomic pile I wondered vaguely why in order to split so minute an object as an atom you need a building about the size of Westminster Abbey. The reason is that the only way to smash atoms on a big scale is to get them to smash one another. This can be done if a sufficient heat is generated within a confined space to make them move fast enough. That is the general principle of the atomic plant. The sun is an atomic plant on rather a large scale. In early days, I am told, Jeans once expressed doubt to Eddington as to whether the sun were really hot enough for this. 'Well, go and find a hotter place!' said Eddington.

It is a cosmic furnace which possesses, as a furnace should, a rim. This is called its gas wall, quite as effective as any wall we could think of, which is held in place by the gravitational pull. How did it come to acquire the sufficient million degrees of heat to promote thermonuclear reactions? Again by the gravitational force. The sun started a comparatively cool volume of gas which gradually became hotter and hotter on account of progressive gravitational contractions which cause great release of energy. When it became so hot that nuclear reactions took place then those same reactions stopped further contraction and a stable sun was the result.

The heat is so great and thus the atomic movement so swift that the outer electrons are stripped off and no atomic 'flesh' impedes the clash of the naked nuclei. Walt Whitman spoke of 'the splendid silent sun'. I have often wondered whether up there it is really silent. Should there not be a series of terrible explosions? This would certainly be so if the material were not just right, so to speak. It is known that hydrogen is the main stuff of the universe, but if there were much lithium as well, then the reaction through lithium into helium would take place in a matter of seconds and the sun would be shattered into bits. If there were much oxygen it would be too slow. George Gamow relates how the great scientist, Hans Bethe, worked the thing out:

'It should not be so difficult after all to find the reaction which would just fit our old sun,' thought Dr Hans Bethe, returning home by train to Cornell from the Washington Conference on Theoretical Physics of 1938, at which he first learned about the importance of nuclear reactions for the production of solar energy; 'I must surely be able to figure it out before dinner!' And taking out a piece of paper, he began to cover it with rows of formulas and numerals, no doubt to the great surprise of his fellow-passengers. One nuclear reaction after another he discarded from the list of possible candidates for the solar life supply; and as the Sun, all unaware of the trouble it was causing, began to sink slowly under the horizon, the problem was still unsolved. But Hans Bethe is not the man to miss a good meal simply because of some difficulties with the Sun, and redoubling his efforts he had the correct answer at the very moment when the passing dining-car steward announced the first call for dinner. Simultaneously with Bethe, the same thermonuclear process for the Sun was proposed in Germany by Dr Carl von Wiezsacker, who was also the first to recognize the importance of cyclic nuclear reactions for the problems of solar energy production.[5]

What it all amounts to is that hydrogen is transformed into helium, not directly but through a cyclic sequence in which carbon and nitrogen play the part of catalysts. A catalyst is a formidable word meaning nurse. Carbon and nitrogen nurse hydrogen into helium, and just as nurses in a hospital, surrounded by dying patients and maternity wards, are themselves fresh and often in high spirits, so carbon and nitrogen take no

5. *The Birth and the Death of the Sun.*

harm from the reaction at which they assist. The final result of this chain reaction – a circular movement – is the formation of one helium nucleus; and one gramme of hydrogen transformed into helium will liberate an amount of energy which may be described in heat units as 166 thousand million calories, or in work units as 200 thousand kilowatt hours. The radiation of the sun is maintained by this exchange occurring millions of millions of times a second.

Is then the sun losing bulk? Yes indeed. Radiation carries mass with it. Since a ray of light makes an impact, a strong enough light directed on a man could knock him down as if with a jet of water. Any body emitting radiation is losing bulk – or weight. The sun is losing four million tons a second, or 250 million tons a minute – which, Jeans suggests, is 650 times the rate at which water is streaming over Niagara. He goes on to say that, 'If it has radiated at this rate for the whole three thousand million years or so since the earth came into existence, its total loss of mass would be 400,000 million, million, million tons'. When I first read those words I felt nervous. All that gone already, and four million tons going every second! Thinking how much one ton of coal makes, it didn't seem to me that we could have the sun with us much longer. But then Jeans adds that even so the loss of those 400,000 million million million tons represents only one part in five thousand of its total mass; and that we may say that each square inch of the sun's surface is only losing about a twentieth of an ounce a century. At this I plucked up spirits again.

Even so, if the sun does keep going for about ten billion years, should it not get cooler? No, it seems that it will get hotter before its final period. This should not surprise us. We do not expect our ordinary coal fires to become less hot as the fuel in the grate is consumed. We know that it frequently gets hotter, while if it is fed too unscientifically by a woman (for women do not understand fires) it will be much cooler for some time. The analogous mechanism of the sun is the changing opacity in the body of the gas while helium is built up, But enough of this. Let us come to Light.

CHAPTER II

THE NATURE OF LIGHT

1. What Is Light?

The nature of light is by no means immediately obvious. If, in complete ignorance, we were to consider it during the day we might well think that it is a thing filling space – rather like air. Then, if during that day we were to go into a room and pull down dark blinds over the windows we would be surprised to find that the room no longer contained any of this thing. It was in the room when we pulled down the blinds and it has not escaped through any hole, but it is no longer present as the air is. Therefore it must be something which is continually arriving, and having arrived it must be instantly absorbed or scattered.

We learn the general statement – that light joins us from the sun after travelling ninety-three million miles in eight minutes. Even so it is not a strictly accurate statement. Light does not exist on its own. It is inaccurate to say that it comes to us in the sense that it is inaccurate to say that a pain has been thrown at me by a boy who has hit me in the face with a stone. The pain is the *result* of the stone travelling through the air and hitting me. If it had not hit me but had broken a window the result would have been different. All I can say is that a force has hit me and has had a certain effect upon me. The sun sends out force, not light or heat. Neither light nor heat has any existence on its own any more than there can be a smell of onions without an onion. The force exists on its own, but it is the results of force which we observe and experience. Various rays go out from the sun, and some of them when they strike an object cause light to scatter and heat to be felt. They must meet the object before light can be seen or heat felt. At night we gaze up into the sky. It is all dark. But beyond the confines of the earth's shadow it should not be dark, it should not be cold; it should be streaming with sunshine, it should be boiling hot – *if* light and heat existed on their own. We can prove this by going up there and being burnt to death in a blaze of sunshine. We need not do so since we have the moon to do it for us. There it hangs, that famous lamp – though itself a barren lightless rock.

Yet sunshine is not all of one piece. It is composed of a considerable number of different rays or waves each of which performs a particular function when it strikes the earth. One wave is interpreted by our minds through the instrument of the eye as light. One is experienced as heat. Another, not the heat-ray, makes us sunburnt. And so on. Each possesses distinctive properties of frequency and length – though by length we do not mean the length of wave but of crest-distance from the preceding one. The frequency of their arrival is thus in accord-ance with their length just as the frequency of Atlantic rollers arriving on the west coast of Ireland is less than that of ripples on a pond in Surrey.

I need not stress that the waves which make it possible for us to experience luminosity and heat seem to be the most precious – but others exist equally vital. Just as heat-waves are incom-petent to excite vision so also there are waves, which though feeble as regards heat and powerless as regards light, are yet of the highest importance on account of their capacity to produce chemical action. In fact they feed us. The whole vegetable world may be considered as a vast Mill receiving its motor-power from the sun. This is so much more than a mere figure of speech that some scientists speak of organic phenomena as altered or differentiated sunshine.

2. The Mills of God

A French cook who was preparing a mouth-watering meal was deeply pained when a friend of Henri Fabre claimed that he knew someone who could produce it all quite easily out of three bottles which he showed the cook. One bottle had nothing in it but air, the other only water, the third a bit of carbon. The cook considered himself a great artist, and he felt affronted by the suggestion that a still greater artist could produce the same wonderful meal out of the contents of those three bottles. Yet so it is. That supreme artist, the green vegetable cell, given a portion of hydrogen, nitrogen, oxygen, and carbon, can, through means of sunlight, build a bacillus, a tree, a mouse, or a man.

Above all, carbon. It seems that carbon is the main staff of life. If we take anything to bits we find that it possesses carbon. The way to take an organic thing to bits is to apply heat to it so that its more volatile particulars fly off. We all know what happens when we do this to bread or sugar or almost anything – we are left with a residue of carbon. So with plants, animals, or ourselves – we can all be reduced to carbon. It is odd that the stuff itself looks so inglorious, for this black substance which smudges the coal-heaver and the chimney-sweep is the insignia of all that is most colourful on earth, responsible for the parade and panoply of the living world no less than for the glittering of the diamond.

We must have carbon. But neither ourselves nor the animals can take it neat – a strong dose of it would poison us. We cannot take it until it has been mixed with something else and is no longer in the form of gas but of concrete substance. The plants on the other hand must take it neat. That carbon is wafted about in the air above their heads. But it is joined with oxygen. The plants do not want the oxygen; yet the two are held together in a tight embrace. They have been welded by the ordeal of fire. How then unburn a stuff forged in the furnace? How pluck from these aerial wanderers the material for the oak and the grass?

Two things are needed: an agent capable of selection, and a force at the service of the agent. The first of these is supplied by that monarch among molecules belonging to the pigment called chlorophyll; and the second by sunlight.

A leaf is riddled with thousands of tiny orifices or mouths which we call stomata. These provide its breathing apparatus. Through these it takes in the same air as we do, though what it wants, as we have seen, is not the oxygen but the carbon. The solution enters and meets with a number of little round bladders filled with a kind of paint, the leafgreen, the chlorophyll. When the sunbeams meet these molecules the work begins, the great work that enthrones the vegetable cell as king, with men and wolves for subjects. First, the beams play upon the leaf and set it tingling all over. The green rays of light come back to our eyes, while the others – chiefly the long wave-lengths – are absorbed into the interior. The chlorophyll molecule responds to them, takes up energy, and thus fortified turns out the oxygen and sets free the carbon. It undoes previous work, as it were, unburning the burnt and de-consuming the consumed. The carbon at once joins itself with the ascending sap to make sugar and starch. All this is done with the swift simplicity of natural miracle; several synthetic steps are taken in less than a second. Building blocks hewn from the primary elements are cast into the crucible, soon to be assembled as grass or tree.

We use a word for this – photosynthesis. And when we use this word 'photosynthesis' what are we saying? We are saying that here is dross; here are bits and pieces; here is matter, cold and lifeless, turned suddenly into living tissue. We are present on the Third Day. This is Genesis. That happened a long time ago we think: yet it happens still, it happens now before our eyes every moment that the beaming blows of light strike on the leaf. These are the mills of God. The solar flood beats on the blades, the flukes fly forward, and the green factory in which the living world is forged, is kept in motion. This is inexhaustible creation. The engine and its agent are wholly unconditioned. For that power is the great sun itself, far removed from earthly weal or woe, unharmed, untouched by any mishaps of the world. And the chlorophyll is largely omnipresent. Whatever man or mouse may choose to do or leave undone, these two agents of the Mover, from everlasting to everlasting, weave living substance from the lifeless air.

From the air. Thence come all things living. We are condensed and consolidated air. We are the offspring of the heavens and the children of light.

We forget this, and at first even doubt its truth as we look round. We see the plants rising from the soil. Are they not laid upon the lap of earth? Do they not lip their mother's milk from down below? They do indeed. They are breast-fed there. But only breast-fed – with mineral water. They must have water though they may sometimes do without soil, as we see in the rootless chlorella carpeting a pond, or when we make seeds grow upon a sponge. And plants can grow from watered soil in darkness; but they are then plant-ghosts and come to nothing, while if air is taken away they perish at once. For every mouth below a thousand mouths must feast above. The rose on the bush, the rose in the cheek cannot bloom without carbon any more than the diamond can shine – which is the purest form of carbon. It is wonderful to observe how plants will appear simply in answer to the presence of light, be it but a lamp. Go into a cave. Venture right into the underground world opened up by the speleologists. This is the realm of darkness. Here little grows. Yet if the speleologists have fixed lanterns here and there upon the walls of the tunnels, and have kept them continuously shining, green plants will grow around them. You can see them in Wookey Hole in Somerset; wherever there is a lamp, then growing in the crevices of the damp rock, we see a cluster of green vegetation. How did they get there? It is hard to say. Light had called them, and they came.

Thus what we see below in glowing forms of life came from above. Imagine the earth before organic life had come. On the one hand, rock: on the other, gas. A bare and grievous land below a poisoned sky fed by the convulsions of vulcanicity and the electric sparks from the storm-clouds. Yet in that angry sky above those hopeless rocks dwelt all that was to be made manifest below; from that tumult and disorder came the organs; from out that fearful riot in the raging flimsy air came all that blooms in plants and moves in creatures, all that smiles and weeps; from those foggy hosts of carboned gas came the vehicle for sensation, thought, conscience, and the word – nay, Plato too sat waiting there, and the body of Christ.

We go forward from that scene through the centuries till we

come to the Carboniferous Forests. That was the era in the history of the earth when trees ruled the world. Their empire comprehended the greater part of all lands. The mind can scarcely seize it. Only the great oceans can compare in size with those forests. They rolled on from horizon to horizon – for once the first speck of protoplasm had been miraculously created the result grew to this, and the trees fed upon a sky saturated with carbonic acid gas many times thicker than it is today. Thus as yet there were no animals nor even insects. We rather shrink at the thought of such a scene; not bare like the desert, not hostile like the sea – but lifeless. We cannot imagine trees whose boughs were birdless, nor gladly think of long-drawn summer days unchoired by bees. Yet so it was. No animal could breathe that acid-poisoned air. It must first be cleansed: atmospheric purification was the primary work of trees. Toiling on through uncalendared centuries, without sound or movement in the speechless glades, they pastured on our poison.

In summary: the plants, whose unbloodied kingdom stretches across the whole world, alone of all living things flourish without hunting and feed without slaughter, simply turning the sky into the tissues of their temples. The sheep consumes grass, the man consumes mutton; neither has yet made any contribution as primary creators, such elaboration is confined to the soundless mills of the green cell in combination with the inexhaustible floods of light poured ceaselessly upon the earth. It is the vegetable which creates the substances, it is the animal which consumes and destroys them. But now a circular movement has been attained which is helpful to plants. The animals permanently pass into the air the acids which the plants need: what the one gives to the atmosphere the other takes. Thus the Circle, thus the Wheel turns forever at its task; the vegetables pupetually decompose the carbonic acid, fixing the carbon and setting free the oxygen, while the animals take the food in the form of prepared and perpetually breathe out that gas: the plants feasting upon the fumes of putrefaction and turning the relics of death into meadows of life, give us green pastures; so that even in our age, riddled as it is with scientific terminology, we can still pay tribute to the simplicity and grandeur of the theme with the rooted ancient words – *All flesh is grass.*

3. Colours

The radiations from the sun do more than light the earth and feed it. They decorate it. Until Newton appeared it was not known that all flowers are black. No one realized that not one single colour belongs to any object intrinsically, and that all objects are black or no-colour. This is contrary to our common sense. We see the yellow daffodil growing up out of the ground, and it seems clear that its colour, the most emphatic thing about it, grows up with it, belongs to it, is it. Yet no; that yellow on the daffodil, that red on the rose was eight minutes ago in the sun.

Before this was discovered in the seventeenth century little was known as to the nature of light and no one had thought of attempting to split it up. It was left for the great Newton to appear with his prism in the dark room. Even today the little experiment still fascinates us. We go into a dark room; we make a neat hole in the blind; a ray of sunshine passes across the room to a screen. Then we place a prism (roughly a piece of glass with at least three sides) in the path of the ray. Instead of the beam passing on – though bent by the glass – we see separate bands of light each of a different colour, the most obvious being red, orange, yellow, green, blue, and violet. A beam of ordinary white light, any thread of it however thin, is a bundle of rays which can be fanned out by a prism because the particular wave-length or scale of each makes it subject to a particular degree of bending. Thus spread out each ray is discerned as a different colour. We close the fan and it is white again. We call this the Spectrum, the word simply meaning spectacle or appearance.

This is a few-centuries-old story now, but few of us are too sophisticated or too dull not to marvel at it. Of course there is no necessity to take a prism and go into a dark room in order to satisfy ourselves that colour exists in white light. We know that all we need is a garden, a garden-hose, and a sunny day. Then we turn on the hose, arranging the nozzle so that the water comes out as a fine spray, and stand with our back to the sun – and there in front of us is a rainbow. The spray has done the work of the prism. It gives me satisfaction to conduct this small research in the only laboratory I possess, the open air. It is pleasant to be able to say – I can pluck colours from the empty air, and set them as an archway over the rose bushes.

Failing a garden and hose, we can simply use a spider's web. When the sunshine is slanting at the end of the day I have often come upon the nets spreading out a good deal of the spectrum. A single dewdrop or raindrop hanging from a twig can be seen throwing back one colour and then another as we change our position. Failing these, we all have eyelashes and noses. They make as good equipment as any. If we sit down and turn our faces to the sun and then screw up our noses while half-closing the eyes (thus looking hideous to a spectator), we shall find that the very fine hairs on the bridge of the nose, together with our eyelashes, will stretch out most of the spectrum.

Thus, then, the waves of light fall upon the objects of the world. What happens depends upon the constitution of the object. One will absorb all the waves, and we get a black object; another will reject them all, and we get a white object; another will accept some and reject others, and we get a vari-coloured object. The black tulip has accepted all and gives us none; the white rose has accepted none and gives us all; the yellow daffodil has accepted most but gives us one. We might be inclined to suppose at first that the colours which the flower accepts should be colours we shall see. But it is a question of reflection. The yellow daffodil absorbs all the colour-rays except yellow which is sent back to our eyes. Thus when we look out upon the colours of the world we are not seeing those which are taken but those which are left. We must not think that the pure red rose in the garden has chosen that colour as its insignia, for red is the only colour which it has refused and cast away.

Can we satisfy ourselves visually that, though the spectrum exists, there are no colours on earth irrespective of it? Not by simple means, I think. I have sometimes read books, otherwise most enlightening, in which it is actually said that if you take a flower into a dark room you find that it has then no colour. But all we can tell in the dark is that it is not luminous – which is an entirely different thing. Yet if we are prepared to undertake extraordinary means there is a way by which we can satisfy ourselves visually that no colours exist in the objects on their own account. We can do it if we pass steadily away from the light of the sun. If instead of going up in a balloon into the heights of the air we descend in a balloon into the depths of the sea we shall behold objects changing their colour before our eyes as the different waves of light grow weaker. Our companion of course would be William Beebe who is accustomed to journey into the deep dark realms of the ocean in his water-balloon or bathysphere. Descending slowly one day, he tells how when less than fifty feet beneath the surface he happened to glance at a large prawn which he had taken with him. Its colour is, as most people know, an attractive scarlet. 'To my astonishment,' he relates, 'it was no longer scarlet but a deep velvety black.[1] Subsequently he studied the changing colours as he went down to fourteen thousand feet. If we were to see this in quick-motion cinematography we would have a picture of remarkable interest, and would be visually convinced at once that these threads of light are responsible for all the coloured patterns of the world.

The ordinary absorption and rejection of waves is taken into account by all of us in simple ways. Thus our chief reason for playing tennis or cricket in whites is the knowledge that if we played in dark tints we would get too hot. When Piccard took his famous gondola high into the air he found it too hot with a dark covering and too cold with a white one, and was at some pains to achieve a colour that would give him a comfortable temperature. Fortunes have sometimes been made on the turf by gamblers who have realized that on excessively hot days, other things being equal, a white horse has a better chance than a dark one. Mountain cottages which are whitewashed become

1. *Adventuring with Beebe.*

far too cold and a dark colour is essential. At Hiroshima those who dressed in white are said to have been spared the worst burns, while those wearing deep colours sustained the deepest injuries.[2]

The discovery of the spectrum made the analysis of spectra (called spectroscopy) possible in our own day. The composition of the rays gives information concerning the properties of the stars. Since matter and energy are the two sides of the same coin we can learn the nature of the matter by the composition of the energy. We are cut off from the material body of the stars by the barrier of impenetrable space. But though the stars conserve their matter they cast abroad their energy. The very barrier of space is no barrier, for it is massed with messages. If we study the radiations we can learn what sort of atoms emit them. They are the messengers who tell us what is going on millions of miles away. We study their story as if we were reading in a book. Swift in passage they are slow in arriving. Energy leaks from matter and streams through space at 186,000 miles a second. Even so the messengers from the Milky Way which went forth at the childhood of mankind have hardly yet arrived.

The scholars who translate these books of light are the astrophysicists.[3] Others deal with the visible, they deal with the invisible. Others are at home with the things of earth, they are at home with the things of space. They analyse the colour quality of the radiation. It teaches them of what materials, of what molten metals, the sun consists – its iron, its copper, its zinc. Thus from a study of light we learn about the source of light and discover that largely the same bricks have been laid to the foundation of the heavens as of the earth.

2. See L. Cheskin: *Colours: What They Do for Us.*

3. I content myself with merely defining this remarkable science. See Herbert Dingle's brilliant *Modern Astrophysics*, and for easy terms see J. H. Fabre's *The Heavens*.

4. The Wonderland

We can see what the world would look like if the accepted colours were apparent to us. Indeed one of the most interesting ways of making the mystery of light-and-colour real to ourselves is by simply putting a prism to our eyes. Then we come upon it with a flash of surprise. Yet it is noteworthy that not a few of those who know all about the prism and the spectrum (in the famous school way of knowing things) do not know what happens if they simply put a prism to their eyes – and probably they think that they do not possess one. But any piece of smooth glass with three sides will do. Personally I use one of the glass fingers which used to hang from a lampshade.

We look through it and see a very different world. So fantastic is the change that at first we may suspect some trickery in the glass. I looked at a line of washing, chiefly white. Through the glass the things appeared rainbowed. The white towel hanging there had a band of orange at the top with parallel bands below of red, green, yellow, blue, and violet. I turned my attention to an area of earth on which I had made a bonfire, consisting now of odd bits and pieces of sticks, silver paper, burnt heads of tins and so forth, on the edge of which was a flowerpot in a white saucer. Through the glass all was transfigured. The flowerpot stood in a bright rainbow. The tawdry pieces of tin became a pearl of great price. The sticks and flints and ridges of clay and ash were coloured walls surrounding something like the Tower of London done up as a fairy palace. A seagull, happening to alight there just then became nearly a peacock in colour-scheme. A well-to-do thrush whose frontage faced the sunlight achieved a fancy-waistcoat.

I took away the prism from my eyes and saw again the bleak area of rubbish. Yet there was nothing in the glass by way of colour to make this transformation.

I looked again. The twigs of a bush which had no leaves on it were dripping with rainbow. A pear-tree with young leaves fully catching the sunlight became a Christmas tree all lit up with candles, some upright and others upside-down. In this case the 'wax' in many colours gave the light. A breeze shook the tree and all the candles waved about without danger of falling off; and so long as the sun shone the candles would burn and the wax not waste. I would use the same image for the flowerpot in the white saucer on the rubbish heap: it was standing in a fire of coloured flames.

I then looked at some withered dandelions. They had come to life again, as also had the withered head of a daffodil: for in this new land there is no death nor sad last days, nor ruin and degradation. A bit of white rag had been thrown upon the path: when I caught sight of it through the glass I wondered what I was looking at – for no flower could shine more bright. A smudge upon the window from inside a greenhouse looked like the Crown Jewels.

Looking up at the houses in this way I saw that all the rims of all the roofs were blue and violet – most lovely and most strange. The white-walled houses threw back all the spectrum, glittering through the trees; and every window which caught the sun aslant was blazed likewise with the bands. So as not to over-multiply impressions I will merely note further that in this mode the gravel path became a Persian carpet, that the hanging piece of tin-foil which my neighbour had put up to scare the birds looked like strangely coloured fish leaping about, and that a page of my writing became an illuminated manuscript.

The paradox about thus seeing the true colours as received on earth is that they give the impression of not belonging to the objects at all, whereas the colours which really do not belong appear to be absolutely integrated. Those blue-rimmed roofs of the houses look as if they had been superimposed by some artist with singular ideas. Another thing is that the objects are not so clearly defined, they are misted and fogged, their fringes blurred. That white shirt upon the line becomes a shirt-ghost, the violet at its lower end being almost translucent. Things which are particularly hard look soft and insubstantial. I was astonished to see that my garden spade was composed of rainbowed fog, and that my clippers lying on the ground reflecting the sun were made of coloured mist – tools fit for heavenly use maybe, but on earth, what dig with this, what cut with those?

Again I look at the glass: it has nothing in it with which to make these changes – save the power to separate the rays. These colours reveal the reality of the waves that paint the rose and gild the lily.

The earth seen in this way is not often as beautiful as when we see it straight. It is sometimes astonishing, as when the wretched rag which has been cast aside is embroidered with bright colours, or the tawdry tin is crowned with jewels; but often it seems a gaudy world and lacking in definition. Purity of line and form have vanished. The rose is ruined; the moon is bilious; the glorious whiteness of the cloud is turned to sickly sunset. We would not care to dwell in this prismed world; but by looking through the glass we get a partial glimpse of the intricate paths and winding ways of light.

5. Sky Colours

Though the mystery of colour was solved three centuries ago, the reason why there is such constant light around us and why the sky is blue has only been really decided more recently. If the force of light must wait till it strikes something before appearing as luminosity, how are we able to see the sky flooded with light? Should it not be dark save for what it receives by way of reflected light from the earth? – so that on the plain the light above us would be very poor, while in the mountains the crags would shine like the moon. The answer is that the so-called empty air is packed with objects massed in sufficient millions to give us the flood-lit sky. Nowadays we are in no danger of imagining that the invisible is necessarily immaterial. The whole of modern physics is based on the fact, as we have seen, that atoms and molecules while less easy to discern than whales are just as physical.

Of course there is a great deal of fine dust in the air up to five thousand feet. It plays some part in scattering the waves of light. It is generally invisible, yet we have all seen it from time to time in the country when the sky is bright but also blocked with weighty clouds sometimes offering just a small hole through which the sun can pass. This familiar spectacle is always delightful and even symbolic. For then a clear beam pierces the gloom like a reversed searchlight or like the shaft above the audience in a cinema, fixing one place for our attention. I have hung a few such pictures in the academy of my memory. One day while working on a high field in Dorset in a close and thundery atmosphere, a hole in one dark cloud let a sun-shaft through. It slanted down to earth, a white beam caused by the thousand million dust-flakes as reflectors, and rested upon a derelict stable in the vale below. It stayed there fixedly. All else was gathered in the pensive gloom, but the beaming finger continued to point to this one ruined and most wretched place as to some truth or treasure.

Nevertheless the invisible dust has only a limited effect on the total illumination of the sky, or with the fact that it is seen as blue. It was thought for some time that the particles of water-vapour were responsible. So Tyndall thought, but it was later established by Lord Rayleigh that the chief agents are the molecules of the air itself. 'The air molecules,' says Bragg, 'are of course very small, much smaller than the wave-length of light, but the cumulative action of a vast number of minute amounts of scattering of the separate molecules is enough to account for the light that we receive from the blue sky.'[4]

The reason why the sky is seen as blue is to be explained by molecule-power in intercepting wavelengths of light. The molecules sift them out. Some of the waves, being shorter than others, are easier to filter. It seems that the air molecules are the right size to intercept the blue waves and scatter them abroad so that we get an impression of blue. When we look at the far-away hills they appear to be blue, and we build hopes on them, and our hearts are raised. But as we approach they lose their glamour and the crags turn grey and comfortless, for they have not been clothed in their own right but are subject to the fleeting

4. *The Universe of Light.*

falsities of the air. It is not so easy for the molecules to scatter the larger waves just as it is not so easy for a rock in the sea to scatter a large wave as a short one. The short waves that compose the blue portion of the spectrum are more easily turned aside than the red waves, but as evening comes on the sun lies low on the horizon and its rays are obliged to traverse many more miles of atmosphere and much more dust than at noon. Most of the shorter waves are filtered out – rather in the same way as when we go down in the bathysphere with Beebe – while the longer ones, the orange, the ruby, and the rose give us the celebrated colours of the sunset.

This is not to say that we cannot on occasion have a green sun or a blue moon. Both have been seen, but not often. More than one observer, including M. Minnaert, have perceived a bright green sun when an engine emitting clouds of steam has temporarily obscured the orb. In fact M. Minnaert got a blue sun as well as a green one out of this.[5] Such phenomena have been witnessed lasting for hours, and without benefit of steam-engines; while after the famous eruption of Krakatoa when vast quantities of fine volcanic dust were hurled into the highest layers of the atmosphere and did not settle down literally for years, people were able to behold far and wide in the world, not only green and blue moons, but sunsets and sunrises of unexampled splendour.

It is possible to see coloured clouds at high noon; but these are more properly called iridescent clouds and they belong to the same family, as it were, as the halo, the corona, the aureole, the fog-bow, the dew-bow, and other rings all caused by complicated diffraction of light by drops of water and mist, some exhibiting pleasant effects as when the shadowed head of saint or sinner is crowned with glory. The term *Heilgenschein* is reserved for the remarkable aureole of light just above the shadow of our heads when in the early morning we stand on the dewy grass with our back to the slanting sun. A special light gathers round our own heads no matter where we move, though we can discern nothing of the kind surrounding the shadowed

5. See his *Light and Colour in the Open Air*.

heads of our companions on the lawn – a pleasing phenomenon which led Benvenuto Cellini to conclude that the shimmer was the insignia of his genius.

It would seem natural to suppose that the aurorae belong to the same category as coronae and haloes, the most famous being the Aurora Borealis seen at high altitudes in the northern and southern skies, looking like an illuminated supernatural aeroplane. But these effects are said to derive from disturbances on the sun and to be associated with the solar prominences and spots when colossal jets of protons and electrons are sprayed out towards the earth at three thousand miles a second, and eventually hit the atoms and molecules of the upper air, electrifying them and making them glow as ghostly streamers of coloured light or as veils and curtains that have caught fire from some cosmic catastrophe.

6. The Rainbow

Let us get back to the ordinary – which generally turns out to be just as extraordinary as the extraordinary. We have already noted how the rainbow exhibits the spectrum. Yet why should we get an arch? The answer is that we do not: we get part of a circle – the whole of which we can discern from a very high place. Again, why only this one circle? Once more we are mistaken: there are many, though we can see only one or two of them. Even so it is surprising. Granted that the water-drops act as prisms fanning out the waves of light, why do we not see a chaos of colours, a ragged and incoherent conglomeration on the screen? How comes it that we see the neat perfect circle? By what means does a shower of water meeting a shaft of light achieve such form? On what spindle and by what potter are the wheels whorled?

It is done by virtue of those geometrical laws which are our comfort and our stay. If our hearts leap up at the sight of a rainbow in the sky, it is made possible by geometry, that mystery lying at the foundations of Nature, as if it were the first of the tools of God. Thus we are told, as a sort of Rule One of the business, that the angle of reflection is exactly equal to the angle of incidence – as it falls so it bends away. It might *not* have been thus ordered. We might have supposed that it could be inconsistent in its behaviour like a refractory child. But never. This is comforting. And it makes for beauty.

Still, how comes it that we get our circle? Imagine a raindrop the size of a plum. It acts as a detainer for some rays, though others pass straight through it. Those which strike its top and bottom are bent inwards and go to the farther edge of the drop which acts as the wall of a mirror, so that the ray comes back to our eyes. Multiply this process so that it is acting all round the drop; and multiply the drop by myriads, though each is being replaced by another as the rain falls! – and we get an inkling of what happens. Only an inkling, I admit. In seeking to avoid the execrable language of the specialists, which, believe me, frequently fails to convey sense, I am not likely to suggest more than a quarter of the operation, so seemingly complicated and yet with all of Nature's majestic simplicity – though I would prefer to make it appear too simple than too complex. And let us not lose sight of the truth about the rainbow which makes us interested in it in the first place – its beauty. In obedience to these marvellous geometrical laws the arches are set up to the delight of all mankind. Their appeal is greatest in the mountain regions, perhaps especially in that part of the world where the light that never shone on land or sea shines on sea and land – in Ireland. The rainbows that I saw in my youth hang before me still. As I write, I think of one. I had been wandering alone in the Wicklow Mountains during many hours of unbroken rain. The leaden sky looked down upon the level bog and the mountains were nothing worth. As late afternoon came on there was a break in the western sky, the sun came out, and the curtain in the east was lifted to reveal the hills from stony crag to curving lawn. It was still showering over there, and a great archway was set up, one pedestal on each side of a valley. At the foot of one, in the rainbow, I saw illuminated a shepherd boy waving his arms rather as if conducting an orchestra, evidently in communication with his dog, and on the other side I could see some transfigured sheep. A simple theme; a pastoral of old time; not quite departed from us yet – though less familiar than when Wordsworth wrote, 'Shout round me, let me hear thy shouts, thou happy shepherd boy!'

7. Fire

One of the main light-spectacles is when the heat-rays produce fire and we get light that way. We have the bundle of rays contained in the spectrum. Some of these rays do not disclose themselves visibly. Thus the ultra-violet waves which are beyond the visible violet ones make us sunburnt. At the other end beyond the red, are the invisible infra-red rays which produce heat. When they are strong enough they become visible in terms of flame – and that flame will itself give us the colours of the spectrum: it will give us the secondary sunshine which we call fire.

I have never found it easy to grasp what fire and flame really are, but I am unwilling to take them for granted. We all strike matches and light lamps and torches without giving a thought to the mystery of it, and regard those who do think about it as special people – 'scientists'. It does seem to me worth while trying to break down this attitude.

Let us take heat first, and note that while it is proper to speak of heat as existing, it is not proper to speak of cold as existing *per se*. There is no such thing as cold save at 273 degrees below. At that point it would be fair to say: Here is coldness. Otherwise it is a question of much or little heat. Sitting in a room I begin to 'feel chilly'. My body is a heat-container, the temperature being about 98 degrees. If my clothes are not 'warm enough', that is if they fail to keep my heat *in*, then I will be letting some of it out into the less warm atmosphere of the room. So I get into bed. Provided I have some good blankets I will now be able to keep the heat in. I can demonstrate the fact that I am not keeping the cold out but the heat in by substituting a block of ice for myself in the bed. The blankets will not warm it up, and make it melt: they will help to preserve it from melting. The air in the room though 'cold' is a good deal warmer than the ice, so if we put the block under blankets we can prevent that warmth from acting upon it. If we realize that there is no such thing as cold in its own right, we are all likely to be more sensible about clothes (as women are); for certain fabrics which are quite light keep in the heat better than heavy ones which simply make us tired. Probably paper clothes would be best from this point of view. Many a down-and-out man has found that a covering of newspaper at night keeps in the heat better than a blanket.

What is heat itself? It is the result of molecular movement – the temperature being in proportion to the rate at which the molecules are moving. When an object is struck by the rays of the sun its molecules are made to move faster, and it becomes hot. Fair enough. But why should fast-moving molecules give us this sensation of heat? I think that the answer is that we then *feel* the molecules, we feel their vibration. When the molecules of the air are made to move fast they are felt by our nerves and we call the sensation warmth. It is often a pleasant sensation. But if we thrust a hand into a very fast-moving bunch of molecules we draw it back quickly as if we had been bitten. We have been bitten – by the molecules. A burn is a bite, as it were: to be scorched is to be mauled by the molecules. But what is this *flame* which we see?

Take an unlit candle. There is a stick of wax: there is a thread of wick. That is all we see at first. Around it there is air which we do not see, nor any part of it – as yet. The molecules that compose the wax are moving at a certain rate but not fast enough to cause the solid to become a liquid. When, by the application of heat, they are made to go much faster the surface of the wax does become liquid and by the law of capillary rises up through the wick. And then it is that from that wick which stands in the pool we see that an extra thing is now fastened – a little spear not made of metal, not made of liquid, not made of vapour, not made of cloth though it is like a little flapping flag, and if you pass your finger through it quickly it seems to have as much substance as the wind. Why should the hot wax take this form? Let us see. Invisible vapour rises from the wax-pool. Vapour is composed of molecules. Molecules are composed of atoms. Atoms possess their outer electrons. The atoms, on account of the speed at which they are now moving became vulnerable; they may lose their outer electrons and combine with other atoms to form a new thing. They combine with the atoms of oxygen in the air. The combination is attended by an explosion – though silent – and we get a spark, a multitude of sparks which make a flame. That flame, once it gets going, once that chemical reaction, as we say, has taken place, carries on by process of feeding upon the wax and feeding upon the oxygen. But we do not know why this action produces this form which we know as flame any more than we know why the combination of hydrogen and oxygen produces water. All we know is that when atoms emit energy, whether from the sun or the earth, we get light. We can describe Nature but we cannot explain the nature of Nature. I have kept a candle burning in front of my manuscript as I wrote these lines, and the more I gaze at the flame the less I suppose myself to understand the miracle of combination and the mystery of combustion with its translation of matter into the comfort of heat and the glory of light.

We do not always need a lighted match, a flame, to achieve a flame. Water will serve. But not in anyway that is useful to us. Thus we build a large hayrick. Unfortunately we have made an error; we have carried the hay when it was too green and therefore contained a good deal of moisture. Without realizing this we go away and attend to other things. Later on we find that the hayrick has disappeared: it has changed into something else and gone off the field – in the form of fire. No match has been applied to it to set off the ordinary chain-reaction. No specially hot sun has been focused upon it. Nevertheless the molecules have been made to move very fast indeed. The moisture has been compressed and cannot escape in the form of gas, and the confined molecules strike against each other so violently that after a while there has occurred a sudden combination and a transformation into flame – for there was enough oxygen penetrating the interstices of the hay to make this possible.

Modern civilization is built on the efficient control of combustion in all its forms. On the occasions when we fail to control it fire changes from our comfort to our terrible enemy. Thus whole cities have been converted into this useless form and gone off, and we think of the great fire of Rome in A.D. 64 when two-thirds of the city went; of Venice in 1106 when nearly all the town passed; of Oslo in 1624 which had to be replaced by Christiana; of the Great Fire of London in 1666 when the greater part of the city was converted; of Moscow in 1812 thus leading to Napoleon's retreat; of the Houses of Parliament in 1834 and the Royal Exchange in 1838 being totally transmuted; of Chicago in 1871 when Mrs O'Leary's cow kicked over a lantern and transformed a city; and so on, including the West Indian Dock fire of London in 1933 which also caused the river to become chiefly liquid rubber as witnessed by myself who swam out into it opposite Wapping, only to land as evidently a native of darkest Africa.

PART TWO
MAN AND LIGHT

CHAPTER I
OPENING WINDOWS

1. The Coming of Glass

In this Part, 'The Story of the Lamp' and 'Sins against Light' have been omitted. Our excesses need not blind us to our successes. The reader might well be surprised if I were to conclude this part of my programme without mention of that manufacture which with light as its ally makes it possible for us to see through walls as if they did not exist, and in so many ways has given us civilization as we know it today. So I come to glass.

Glass is not a natural compound. Great heat turns sand into a liquid. When that liquid cools it becomes glass. Pure glass is an invisible stone, as it were. We cannot see it. It is not apparent, it is transparent we say. We see through it because the waves of light pass through it. It does not absorb all of them and give us a black stone. It does not reject all of them and give us a white stone. It does not select some of them and give us a coloured stone. The waves pass right through – quite easy, since everything is chiefly made of holes.

But we cannot speak of pure, clear glass as the kind which was made first. Glass has been manufactured for some five thousand years, but for many thousand years no one knew how to make uncoloured glass, something which you did not look *at* but *through*: you always saw through a glass darkly. Its history has been long and fruitful, from the Egyptian beads to the twenty-ton block for the 200-inch telescope. We may note in passing that beads of glass have always exercised fascination, being especially appreciated by the untutored mind. 'When the white man first came to America,' write Rogers and Beard in their brilliant book, *Five Thousand Years of Glass*, 'he found the red man already occupying it and, strange to relate, the Indian entertained the impression that the land was his. If the white man wanted land he ought to pay for it.' The white man, who by the way was not yet 'an American', was deeply pained at this, not to say morally shocked. However, since the Indian was apparently unwilling to give up his inheritance without compensation the question arose as to whether there was anything he might be offered as a price for acres. Glass beads were tried. The Indians were delighted, thinking them a fair exchange for virgin soil. Since they were satisfied with glass beads why waste valuables on them? A bead-mint was set up to the satisfaction of everyone – for a time.

It was not until the rise of that City whose inhabitants by paving their streets with water were to fill their pockets with gold and yet silence the roar of modernity, that glass came into its own to assume lovely shapes and blossom in abundance like flowers; and by the time that Venetian glass had become famous the secret of making it transparent had been discovered.

We still pause in wonder before a thin wall which we can see through. 'Windows are openings in the wall to admit light and air,' says the *Encyclopaedia Britannica*. I had always thought that windows were light-openings in the wall by virtue of glass, thus admitting light but not rain, snow, or tigers. Yet of course the *E.B.* is right: windows did not primarily have glass and were simply openings, often being another description of the main door. Sometimes you closed your window with wood or cloth; the idea of filling the space with glass came quite late, and ordinary house window-panes were such a luxury even in the America of 1829 that we learn that the Rev Mr Higginson of Salem advised all newcomers to bring their own panes with them. In Europe there were long periods when glass windows were not wanted at all. For ages the intermittent wars caused houses to take on the aspect of forts in which the windows were

mere slits in the wall. However, eventually real window panes
began to come in until we reach modern times when some
houses have whole sides made of them. In England it took some
time before they assumed generous proportions, because between 1696 and 1831 the government pursued the good idea of
taxing light and air. It grieved the more progressive business-
men in the government to see such an excellent commodity as
the sun being had for nothing. The difficulty was how to get
people to pay for it. The problem was solved by a charge in
terms of a window-tax. This policy met with sustained support
for one hundred and thirty-five years from the best minds who
considered that the fevers and plagues, which resulted from the
insanitary conditions thus imposed, could scarcely be set against
the advantages of the revenue received. A considerable number
of cottages in Britain are still standing as a relic of those days.
In the end, however, light was permitted to enter free of charge,
and the abundant blessings of big glass windows have been
enjoyed for a hundred years – which is as good a way as any of
worshipping the sun and paying tribute to light.

2. Glass and Civilization

The pure pane of glass through which light can easily pierce
made possible the invention of our one happy prison. In a glass-
house we can imprison heat. Having got in it cannot get out,
and remains there for some time even when the sun has ceased
to shine. My own appreciation of this is far from academic. I
am aware how profitable it is for the growing of vegetables,
fruit, and flowers; but I use one as a study and am able, on
occasion, to sit down at a table in April when in actuality it is
December or January. I have written about the sun
scientifically in this book, but I have no desire to conceal the
fact that I have always, and from a very early age, had a simple
adoration of the sun as we know it in our part of the world. I
cannot say with Wordsworth:

> ... already I began
> To love the sun; a boy I loved the sun,
> Not as I since have loved him, as a pledge
> And surety of our earthly life, a light
> Which we behold and feel we are alive;
> Nor for his bounty to so many worlds –
> But for this cause, that I had seen him lay
> His beauty on the morning hills.

I cannot quite say that, for though I did see him lay his beauty on the Wicklow Mountains, I always loved the sun simply because it made me happy. The warmth without gave and gives me warmth within – even light within. To receive it in March or November from behind glass (with a cold wind outside) prompts me to draw up a list of things I could do without if I could have this thing. Why cannot the heat get out through the glass if it has got in? The answer is what we should expect: heat waves are a force, and the force is largely spent in getting in.

So many things followed the introduction of transparent glass that it has been fairly claimed that a list of them in terms of usefulness, knowledge, comfort, culture, and entertainment might serve as a description of modern civilization. Indeed this is so much the case that we have ceased to realize it. To take a single example, it scarcely occurs to us that the Romans never saw themselves in the mirror. The best that Cleopatra could make use of was a shiny surface. This was kind to the plain, but today we suffer from an excess of mirrors, forever being confronted with our faces. They have exercised a great fascination for people, and a history of mirrors would draw attention to the mirror-epidemics which have broken out in societies again and again. In the field of literature we got Alice. Sociologists even claim that it promoted biography. Thus Lewis Mumford makes the curious suggestion that 'the use of the mirror signalled the beginning of introspective biography in the modern style: that is, not as a means of edification but as a picture of the self, its depth, its mysteries, its inner dimensions.' Though he fails to support this with enough evidence, it is undoubtedly true that the bestowal of *depth* is one of the chief gifts of the mirror. We see deeper into reality, and discern more in the commonplace than before. This is partly due to the fact that the mirror is a *frame*. If we frame anything we see it, not only more clearly but with fresh vision; and this, I think, is why the framework of a novel is so important, for if that frame is good, then, in spite of a great deal of coincidence and unlikely re-meetings and so on, we get a greater concentration upon essential reality. Even when, standing back within a room, we look through a doorway or a window we see the section of the commonplace thus cut off, with new eyes. A mirror frames us back our own scene, somehow making it more mysterious and aloof and in a sphere of beauty[1] (note how the reflection in a motor-car's mirror, when going through an autumn-tinted tree-lined road, say, seems more significant than the 'straight' scene, however appealing). So fascinating is the mystery of mirrored reality, especially when presented the more elusively and re-motely by water or copper, that Jacob Böhme declared that it was among the polished pots of his kitchen that he received intimation of the secret light of the Universe. We can scarcely hold it against Narcissus that he formed so favourable an impression of his mirrored image; and we may agree that the celebrated dog who preferred the reflection of his bone in the water to the bone itself, had the root of the matter in him.

Mirrors, though comparatively recent, are not modern any more than hot-houses. Nor are spectacles. In the thirteenth century Roger Bacon had already noted the properties of a dew-drop, how it acted as a lens bending the rays and magnifying or distorting the object. Lenses of glass were in use in the fifteenth century as eye-strengtheners, and at the end of it concave lenses came in by means of which the distorting lens of an imperfect eye was corrected by an opposite distortion (two wrongs thus making a right). This greatly added to the reading life, and with the invention of printing threw open those intellectual windows

1. Note the mirror in Jan Van Eyck's famous 'Jan Arnolfini and his wife'. Thus in this picture the artist makes a double aesthetic assault.

which we call the Revival of Learning. We enter modern times when the possibilities of the magnifying-glass became fully realized. We know what the camera has meant to us – even allowing us, through means of the rays we call X, to photograph our insides and see how things are getting on there. Everyone has his own idea as to what the cinema is: personally I see it as a window through which we can look across to the farthest ends of the world.

3. The Window of the Telescope

There is no turning point in the alteration of man's perspective so complete as when in 1610 Galileo, after working night and day on his fifth telescope which made things thirty times nearer and a thousand times larger, viewed the heavens and announced what he saw in his *Siderius Nuncius.*

Today we brood upon the vastness of space. Before 1610 they saw just the same as we see when we regard the sky on a clear night without a telescope. But they did not conceive what we conceive. Evidently the prestige of the old cosmology was so great that all those stars worried them no more than if they were decorations upon a ceiling to the world. Then suddenly the idea of there being any kind of ceiling or even boundary was taken away and they were confronted with an infinite vastness. That was depressing enough. But the new findings went much further, not only refusing the assumed centrality of our earth but presupposing the existence of other earths such as our own. 'A new philosophy arrests the Sunne,' declared Donne, 'And bids the passive earth about it runne.' That was bad: but Galileo's discovery of some new planets introducing the idea of a plurality of worlds was theologically unthinkable. The powerful Church could not possibly allow it. Christ died for *this* world. A plurality would offer shocking obstacles to Faith. If God was to be conceived as the One and Only then He must be responsible for the creation of all the universe. If there were other worlds like ours in which He had placed imperfect beings in equal need of salvation, then – how could they be saved? The Church must instantly suppress such ideas. And just as Roger Bacon at an earlier date was thrown into prison for fourteen years because he insisted that the rainbow was not a special sign from heaven but a natural phenomenon, so Galileo must recant before the Inquisition, lest 'our whole beautiful system fall to the ground.' He did so, and spent the rest of his life a captive in his own house. His recantation was not thought strange at the time, but as the centuries passed his outrageous betrayal became such an unbearable blemish upon his name that future biographers invented his subsequent whisper after the denial: 'It is true all the same.'

However, since zealots have always been finally powerless against the majesty of truth the telescope itself could not be suppressed, and the sheer effect of its sense-perception, so much more powerful than any argument, shattered the old cosmology for ever. The homely conception of the universe was abandoned and men were compelled to contemplate the enormities of interstellar space. The Elizabethans had been perfectly happy with Aristotle and Ptolemy. The imagination of the greatest master of words the world has known was never stirred by Space. Time ruled Shakespeare like a king: in his day it was even more of a tyrant than in ours, for death knocked at the door so soon in those plagued and fevered times. The plays are riddled with Time and his sickle – but of Space there is nothing. And instead of astronomy just a little astrology in *King Lear*. Then the heavens fell. Now it was open to some new great poet to grapple with the new conception. There was such a man. He had the imagination. He had the language. He had the art – even to encompass the illimitable. He did not do so. Instead he gave us the great ruin known as *Paradise Lost*. He gave us Genesis and the Garden of Eden in the very age that had opened up interstellar space. It was like building a cathedral to be bombed as soon as it was finished.

Milton thought he could have it both ways: the old story told against a background of the new cosmology. It would not work out. He could not afford even a minimum of scientific exactitude, and in any case the whole cast of his mind and educational upbringing was anti-scientific. But neither could he afford vagueness and evasion since these are the enemies of imagery. Heaven is up there somewhere; hell is down there somewhere; the earth is over there somewhere: perhaps that amount of vagueness would not have mattered any more than the creation of Light on the First Day and of the Sun on the Fourth Day as retold in Book VII, if the epic had been written in an earlier age. But Milton who was born in 1608 had been to Italy and had visited Galileo and looked through his telescope. That was naturally a tremendous experience. His imagination took wing. In Book II Satan leaves Hell to visit the garden of Eden. He stands on the brink, 'Pondering his voyage; for no narrow frith he had to cross.' He found himself faced – *with the new cosmos.* It was terrible to contemplate. It was an illimitable waste without bound or dimension in which time and place were lost and where eldest night and chaos, ancestors of nature, hold eternal anarchy. It was a boiling and abortive gulf. It was a dark, unbottomed, and infinite abyss. It was an uncouth way, a vast abrupt, a profound void, a wide womb of uncreated and inessential night threatening the traveller with utter loss of being. Yet Satan takes the plunge. Soon he meets 'a vast vacuity' causing him to fall ten thousand fathoms in the wrong direction. He would be falling still, says Milton suddenly in the manner of a modern astronomer, 'and to this hour down had been falling', had he not 'by ill chance' met with 'a tumultuous cloud' which threw him in the right direction for Eden.

We need a universe which the kind of God depicted could create, we need a system offering the possibility of justifying the ways of God to man – that is we need definiteness and limitation. But the epic is definite and indefinite, it is limited and unboundaried and riddled with anomalies such as the presence of clouds where there could be no clouds, and air where there could be no air, and water where there could be no water, and even bogs where there could be no bogs, and on one occasion there is speculation as to how Satan could have shown Christ all the kingdoms of the world and the glory thereof without using a

telescope.[2] Milton, whose mind moved with splendour in the
field of classical allusion, did not consider that a blending of the
natural with the unnatural and the physically ordinary with
the impossible, mattered much so long as he got good effects;
and when in Book VIII Adam remonstrates with Raphael on the
grossly uneconomical manner in which the other stars admin-
ister to this one 'punctual spot', our little earth, he is told not to
bother about it. This kind of thing does not trouble us when we
read the poem in youth, for we surrender to its marvellous
rhetoric and organ music; but later on we want the system to
make sense, and if we do not go so far as Swift who implied that
it would be just as sensible to hold that the universe is a large
suit of clothes which invests everything, the earth by the air, the
air by the stars, and the stars by the *primum mobile* we are
inclined to lay down this Epic and perhaps take up *The Prelude*
which treats of 'this world which is the world of all of us, where
we find our happiness or not at all.' *Paradise Lost* stands in
literary history like some great Ruin on the plain, which not
even in the past was ever inhabited, its great halls abandoned
and bare, its glorious stained-glass windows long since shattered
by the impious.

4. The Window of the Microscope

When in the seventeenth century the windows were opened and
Milton's generation heard 'a shout that tore Hell's concave,
and beyond Frighted the reign of Chaos and old Night', it was
not only sudden and unexpected but alarming, save for those
strong souls who like Milton himself really exulted in the wild
freedom of that frantic thought, and like Leibnitz beheld
therein 'overflowing Benignity and Divine Super-abundance'.
The world opened up by the microscope, which was in full use
by about 1660, was less unexpected. With remarkable fore-
knowledge and exactitude of prophecy Francis Bacon had prepared men's minds in his *Novum Organum* for the secrets 'still

2. Actually this comes in *Paradise Regained*.

laid up in the womb of nature' which would be revealed by new instruments in the years to come, and in *The New Atlantis* there are 'glasses and Meanes to see the small and Minute Bodies' and to inspect the structure 'of small insects and the flaws in Gemmes', though Bacon himself did not live to see his prophecy fulfilled.

The revelations of the microscope were in general received with wonder and joy. There are always some men temperamentally unable to face new prospects and who dwell perpetually in the shadow of the 'Night of the Soul'. Thus Pascal could not bear the new cosmology. 'The eternal silence of these infinite spaces frightens me,' he said. 'When I see the blindness and wretchedness of man,' he cried.

when I regard the whole silent universe, and man without light, left to himself, and, as it were, lost in this corner of the universe, without knowing who has put him there, what he has come to do, what will become of him at death, and incapable of all knowledge, I become terrified, like a man who should be carried in his sleep to a dreadful desert island, and should awake without knowing where he is, and without means of escape. And thereupon I wonder how people in a condition so wretched do not fall into despair.

I would not for a moment question the ultimate profundity of those who grieve like this, nor deny 'the ineffaceable, sad birthmark in the brow of man' as Melville read it, adding that 'the mortal man who hath more of joy than sorrow in him, that man cannot be true – not true, or undeveloped'. Yet at all times men have sought for signs and answers in the glass. In discerning the principle of beauty in all natural forms they have felt it as a promise and received it as a benediction. Not so with Pascal. Having surveyed the cosmic order he turned to 'the prodigy equally astonishing' revealed by the microscope. It also frightened him. The flea now seemed to him a fiend, 'its limbs with joints, veins in these limbs, blood in the veins, humours in this blood, globules in these humours, gases in these globules'. It was another abyss, 'the immensity of nature in the compass of this abreviation of an atom'.

But this was far from being the general response to the microscope when glass allied with light gave us eyes to pierce into the minutiae of creation. It is true that its novelty sometimes alarmed simple people such as the inhabitants of a Tyrolese village who refused Christian burial to a certain man who had died with a fly in his pocket-lens, on the ground that he possessed on his person 'a devil shut up in glass'. But for the most part its revelations were hailed with gladness and were responsible for many a charming sentence from Leeuwenhoeck and Power, from Hooke and Browne and many others. They felt elation – as who does not? – when they traced the delicate fingers of design printed upon the tiniest envelopes of life. Henry Power marvelled to see on the butterfly 'the very Streaks of the Celestial pencil that drew them',[3] while the assistants of Leeuwenhoeck on drawing a flea and seeing 'the *little wheels* on the animals in *swift rotation*', continually cried out, 'O that one could ever depict so wonderful a motion!'[4] And just as to Sir Thomas Browne it seemed that 'in these narrow Engines there is more curious Mathematicks; and the Civility of these little Citizens more neatly sets forth the Wisdom of their Maker'[5] than is to be discerned in the motions of an eagle, so also Robert Hooke when gazing upon the seeds of thyme and noticing how well they are protected from outward dangers, observes in language as felicitous as the illustrations to his wonderful *Micrographia* that Nature 'as if she would from the ornaments wherewith she hath deckt these Cabinets, hints to us that in them she has laid up her Jewels and Masterpieces' and that 'the Creator may in these characters have written and engraved many of His most mysterious designs and Counsels'.

This discovery of the minutiae preceded by at least a century the discovery of wild nature in mountains and waterfalls as means to experience the divine. Indeed Henry Baker suggested that it were better to look at a mite than at an elephant, while Robert Boyle thought that 'wonder dwells not so much on

3. *Experimental Philosophy* (1663).
4. Marjorie Nicolson: *The Microscope and English Imagination*.
5. *Religio Medici*.

Nature's Clocks than on her Watches'; and it was not only Blake who beheld the universe in a grain of sand but Traherne who declared that 'you never enjoy the world aright till you see how a sand exhibiteth the wisdom and power of God'.[6] This was a new attitude. Generally speaking, in the seventeenth century before the arrival of the microscope Nature was thought of as ragged and uncouth over against Art which was so elegant and finished. With the coming of the microscope the tables were turned: it was Art which seemed rude and Nature that was polished. Thus when Henry Baker looked at artefacts through the microscope he discerned only a concealment of deformity and an imposition upon our want of sight, whereas he claimed that in Nature the more we look at the least and meanest of her productions, flea or fly or louse or mite, nothing can be found but beauty and perfection; and when he examined the particles of matter composing salts and saline substances set at liberty by dissolution, and beheld the order in terms of rhombs, pyramids, pentagons, hexagons, and octagons in mathematical exactitude, 'moving in Rank and File obedient to unalterable laws in determined figures', he saw 'Almighty Wisdom and Power'.[7] John Wilkins declared that 'whatever is Natural doth by that appear adorned with all imaginable Elegance and Beauty', as against the 'rude bungling and deformed work' of men, so that the finest needle seems but blunt when compared with 'the inimitable Gildings and Embroideries in the smallest Seeds or Head and Eye of a fly' such as issue from 'nature's forge and furnace',[8] while to George Adams the Younger (the Elder wrote in the same strain) the contrast was humiliating, and when he saw closely the many fibres of a flower he was enchanted, adding that 'the whole substance presents a celestial radiance in its colouring, with a richness superior to silver or gold, as if it were intended for the Cloathing of an Angel'.[9] Yet perhaps the strongest impression made was that of an endless fund of life to be

6. *Centuries of Meditations*, No. 27.
7. *The Microscope Made Easy.*
8. *The Principles and Duties of Natural Religion.*
9. *Essays on the Microscope.*

discovered everywhere. Addison's tiny dream-visitor in the world of the microscope beheld 'millions of species' on a green leaf, and became lost in a forest of trees growing in the cup of an acorn;[10] while there is a curious up-to-dateness in the declaration of Fontenelle that 'solid Bodies are nothing but an immense swarm of imperceptible Animals' and that 'everything is animated, and the stones upon Salisbury Plain are as much alive as a hive of bees'.[11]

Even so it was not the delight of the poets, nor the conclusions of the philosophers, nor the amusement of the satirists, from Swift with his Gulliver and his famous quatrain on the flea, to those who made play with the 'philosophic girls' who preferred microscopes to lovers, which made this instrument so important in laying the foundations of modern civilization. It was its effect on medicine. In 1674, Antony van Leeuwenhoeck when passing a lake decided to take a pail of water from it and examine it through his microscope. He saw creatures a thousand times smaller than the smallest ones he had seen upon the rind of cheese or in wheaten flour or mould. Then, as he told the Royal Society in his Letters, he sampled a pail of rain-water and found creatures 'more than ten thousand times smaller than the ordinary animalcules'. He was delighted with the perfection of these tiny creatures also; he felt abashed before the Creator to behold structures 'so delightsome and wondrous'; the whole water seemed 'to be alive with these multifarious animalcules. This was for me, among all the marvels that I have discovered in Nature, the most marvellous of all'. He had discovered bacteria and protozoa. He had opened the door of microbiology and bacteriology. He had thus laid the foundations for the observation of the circulation of the blood. He had paved the way for the introduction of all the antitoxins which rescue us from so many plagues and fevers. He was responsible for the final marvels of Fleming's penicillin no less than for the inoculations of Pasteur.

Still, I am unwilling to close my account of glass and light on

10. *Tatler*, No. 119, Jan. 12, 1709.
11. *A Plurality of Worlds, pp.* 89–90.

this note. I have kept for the last our happiest victory in this kind. It is entirely aesthetic and spiritual. It is not the transparent variety I have in mind, but stained-glass. Man has never guided light to better purpose than with the stained-glass window in church or cathedral. A beautiful specimen of this art in a place of worship, not overlighted by plain windows, works upon us. It may be a cathedral in the great city; better still, a small church enfolded in the English hills, sequestered and hung with history. At any time we can enter into the seclusion of the holy place, and wrapped in the solitude that is not loneliness gaze up at the window where light turned to colour receives light again; and if the beams of the setting sun fall upon it so that the colour of the ray works with the colour in the waves – a double effect, a multiplied miracle – then truly physics joins with metaphysics, and the sombre radiance of that melancholy light belongs to the kingdom of heaven.

CHAPTER II

SUN WORSHIP

1. Sun-Gods

As I passed along a country lane recently on a winter late afternoon I turned a corner to find that a big red ball had got stuck in the twigged network of a hawthorn tree. It was too large for children to play with if I took it down, but might serve. I did not find it really natural to accept the fact that it was not in the tree but over ninety million miles away.

Sometimes a simple little experience of that sort gets us closer to the mystery of things than our present extensive intellectual knowledge. Today we know a great deal about the unity of, and our own unity with natural phenomena – but we do not feel it. Our ancestors felt it – though they did not know much about it.

It is really a pleasure for us to contemplate the cosmology of earlier men. In modern times we interpret the heavens not as the eye sees it but as the mind knows it. As their eyes saw it so their minds conceived it. Over their heads was the blue, white-scarfed roof of the world at night tastefully hung with lanterns, the moon a radiant face about ten miles off, while above the roof lay heaven, where you went after death and had a good time. It was not too difficult to feel at home, even if fearful, in such a universe. The sun was no lonely, lofty castaway millions of miles off, but a Being who would be glad to be offered a drink to quench his thirst as he completed his exhausting journey across the sky. He was a god: and a god in early days was always regarded as remarkably human. Thus Phoebus Apollo, the most famous of all sun-gods, was so clearly conceived in personal terms by his devotees that they thought him to be jealous enough to cause the ears of King Midas to lengthen till they resembled those of an ass and had Marsyas flayed alive for

69

saying that he played the flute better than himself; and later when Anaxagoras not only denied the idea of the sun's divinity, the idea of its personality, and even the idea that it was a Universal Eye which beheld everything, and declared that it was merely a mass of hot iron about the size of the Peloponnesus, he met with the greatest indignation and escaped death only through the intervention of Pericles. The Cherokee Indians thought that the sun hated the people of the earth because they never looked straight at her without screwing up their faces – 'My grandchildren,' she is represented as saying, 'are ugly; they grin all over their faces when they look at me, but for the Moon they smile.'

Those who could watch the sun daily rising from the sea took it as a white bull, and for thousands of years the Kings of Egypt delighted to style themselves Ra, the mighty bull; while the phenomenon of it sinking into the sea made them think that it was swallowed by a monster who in the morning disgorged its prey into the Eastern sky – a myth said to be responsible later on for the story of Jonah and even of Little Red Riding Hood. It is not surprising to learn that in some quarters the people, expecting it to make a hissing noise as it sank into the water, claimed, in the manner of eye-witnesses and ear-witnesses the world over in all ages, that they did actually hear that sound. The Aegean islanders believed that every evening the sun returned to his kingdom in the underworld where his mother waited for him with forty loaves. If they were not ready he would eat his entire family. On rising red from the sea the islanders said: 'He has eaten his mother.'

When we survey the sun-legends of early man and wish to understand them, we must be careful not to think in our own consequent terms. It was quite possible to put out the sun if you wanted to make mischief, like that poor Indian boy who finding people uncharitable to him threatened to shoot the sun, and did so, and the light went out at once, and the whole world became dark. It was not always assumed that you could have a sun unless you took the trouble to get one: thus Coyate-Man is represented as going on a journey and finding a land flowering with sunshine. He offers to buy it, but they will not sell. So he resolves to steal it, and does so, though it was carefully guarded by a turtle. The Apache myth held that when their ancestors wanted a better system of lighting they simply devised a sun, while the North Americans believed that either a wolf or a raven or a hawk had procured one for them. It could perform simple tasks for you: since it is obviously the sun who sends down hailstones, it could easily send you a new tooth. Thus an Arab boy, on his tooth falling out, would take it between finger and thumb and throw it towards the sun saying: 'Give me a better one for it.' The sun is seen to travel backwards and forwards, so why should it not give one's absent husband a lift home on the return journey, complying in compassion to the plaint: 'May the grief that my abscence causes him make him weep, may the grief that my abscence causes him make him lament, may the grief that my absence causes him make him break the obstacles that part us and bring him back to me at sunrise.'[1]

2. The Sun as Saviour

Where we would say that the sun is rising high in the heavens, they said: 'The sun has harnessed the horses for his journey.' Swift horses were associated in the solar mythology of the Indians, the Persians, the Hebrews, and the Greeks. Sometimes it was given wings and thought of as a bird – 'The bird of day is weary and has fallen into the sea.' The North Americans saw it as a hare. The Greeks wove the most charming fancies around the expedition. Seeing in the flashing rays spirited and fiery steeds harnessed with gleaming trappings and burnished reins they annually dedicated to him a carriage and four horses, and flung them into the sea for his use, since his old ones would be worn out. It was sometimes considered he went too fast, and wishing to retard his progress certain races used a net to catch him, or summoned a strong man to imprison him in a tower, or

1. J. G. Frazer: *The Magic Art*, I. p. 166.

in the manner of Joshua frankly commanded him to stand still. Indeed, they were far from thinking of the sun always as a very powerful god. To many it seemed strange that he should mount the sky each morning with absolute regularity and pursue the same path. They thought he must be compelled to do so. Once he may have done what he liked until caught in a trap and beaten into submission, bound with cords, held fast by eight hundred thousand gods who have ever since retained him. The Inca of Peru denied the sun's deity as supreme, on the ground that his course was so circumscribed. 'If he were free,' he declared, 'He would visit other parts of the heavens where he had never been; but as he follows one path he must be tied like a beast.'

Yet the most pleasing legends spring from the conception of the sun as saviour and servant of mankind. It has a task which it must perform. It has a path to travel from which it must not deviate. It has ever before it a life of toil from which it cannot swerve. Thus the heroic figure of Hercules, still our symbol for the greatest deeds of endurance and strength, was shaped by the imagination of mankind. 'Nowhere,' says Cox, 'is the unutterable toil and scanty reward of the sun brought out so prominently as in the whole legend, or rather in the mass of unconnected legends which is gathered round the person of Herakles.' Interpreted in any other light his adventures are rather tiresome but considered 'as the luminary that gives light to the world, as the god who impregnates all nature with his fertilizing rays, every part of the legend teems with animation and beauty, and is marked by a pleasing and perfect harmony'.[2] in a more simple form the figure of Sisyphus who was condemned to spend his days laboriously rolling a great stone to the top of a hill, after which it rolled down again, signifies the solar sphere gradually mounting its zenith each day, only to slip down again. We must add the agonies of Ixion and the tortures of Tantalus, the one bound for life to a four-spoked ever-revolving wheel, the other with his head above water, dying of

2. Anthon's *Classical Dictionary*

thirst yet tantalizingly unable to drink, as further examples of solar symbolism.

Since primitive man always strove to interpret natural things in terms of the familiar language of daily existence and to attribute a human agency to manifestations of physical laws, the part which night and darkness played in the making of myths was necessarily very great. 'In the thought of these early ages,' says Cox again,

the sun was the child of night and darkness, the dawn came before he was born, and died as he rose in the heavens. He strangled the serpents of the night, he went forth like a bridegroom out of his chamber and like a giant to run his course. He had to do battle with clouds and storms, sometimes his light grew dim under their gloomy veil and the children of men shuddered at the wrath of the hidden sun. His course might be brilliant and beneficent, or gloomy, sullen, and capricious. He might be a warrior, a friend, or a destroyer. The rays of the sun were changed into golden hair, into spears and lances and robes of light.

In the night another luminary appeared, the moon, and it was natural that the sun should be conceived as pursuing the moon across the sky, the coy moon for ever flying, yet the pair meeting in matrimonial embrace in the interval between the old and new moon; and this was thought to be a good time for human marriages, for on the principle of sympathetic magic when the sun was wedded in the sky men and women should be wedded on earth. Some thought of the sun and moon as brother and sister or husband and wife from the beginning, and the stars as their children. Other legends suggested that the moon agreed to eat her children for the benefit of mankind who could not bear so much heat and brightness, and that the sun devours her children at dawn. It seemed also that the sun often bit the moon, either angrily or amorously, taking a neat curved slice out of it as from a cake, occasionally leaving only a very small piece. One legend tells how the Sun, the Moon, and Pole Star were all suitors for the hand of a beautiful maiden hatched from a goose's egg. The girl objected to the Moon as unstable, with a face which sometimes narrowed in an unpleasant manner, and with a bad habit of roving about all night and remaining idle at home all day. She did not regard the Sun with much more favour since he was the cause of too much heat in summer, cold in winter, and uncertain weather conditions. She chose the Pole Star who always came home punctually.

The Norse myths display just as many sun-heroes who accomplish impossible tasks and vanquish formidable foes. The sun is Odin's eye, Balder's countenance, Heimdal's need – 'And still she rides, the beaming maid, from morn to night.' In the Finnish legends we find the sun as a lamp illuminating the halls of Vanna Issa, the Supreme Deity, and entrusted by him to the care of two immortal servants, a youth and a maiden – Dawn and Evening Twilight. Thus, in the words of this charming creed, the Father says: 'Unto thee my daughter I entrust the Sun. Extinguish him and hide him lest he come to harm. And unto thee my son I entrust the duty to rekindle him for a new course. On no day must the light be absent from the arch of heaven.' There is melancholy as well as poetry in the touchingly personal tale.

In the winter he resteth a great while, but in summer his repose is short, and Evening Twilight gives up the dying light into the very hands of Dawn who straightaway kindles it into new life. At such times they each take one look deep into the other's dark brown eyes; they press each other's hands, and their lips touch. Once a year only for the space of four weeks they come together at midnight. The Evening Twilight layeth the dying light into the hands of Dawn, and the cheeks of Evening Twilight redden, and the rosy redness is mirrored in the sky till Dawn rekindles the light.[3]

3. W. T. Olcott: *Sun Lore in All Ages*

3. Solar Relics

'In the childhood of mankind,' says Paley, 'the daily death of the sun was regarded as a reality.' It is a little difficult to credit this, seeing that it happened every twenty-four hours. But it is clear enough that its efforts were a constant topic of anxiety, and it is certain that once a year it alarmed them terribly: it began to fail in autumn and to fall into increasing depression as winter drew on. Older than all history and all written records has been the fear that took hold of the people at this decay of their god of light and warmth. Was he dying, or would he revive and reappear? – that was the awful question. There were no almanacs, no calendars, no systematic ordering of time into seasons and years, no memory about what happened last year – all they knew was that their chief source of comfort and life was failing. What was happening during the short days which we know as the end of the year? The god had fallen upon evil times. He had come under the malign influence of the Serpent and the Scorpion. Delilah, the queen of night, had shorn his hair (Samson is derived from *Shemesh*, the sun); Typhon, the prince of darkness had betrayed him; the dreadful Boar had wounded him; Herakles was perhaps himself being slain at last. Would he grow weaker and finally succumb? 'We can imagine the anxiety with which those early men and women watched for the first indication of a lengthening day; and the universal joy when the Priest (the representative of primitive science) having made some simple observations, announced from the Temple steps that the day *was* lengthening – that the Sun was really born again to a new and glorious career.'[4] That would have been roughly on 25 December.

The time came at last when natural phenomena were no longer deified, and one invisible god took the place of the many visible ones. But numerous festivals and rites of solar origin are with us still, entwined into the fabric of modern civilization. The Roman festival of the winter solstice, celebrated on 25 De-

4. Edward Carpenter: *Pagan and Christian Creeds.*

cember in honour of the sun-god Mithra, known as the 'Birthday of the Unconquered Sun' was introduced into the Christian Church in the fourth century as a festival in honour of the birth of Jesus Christ. Augustus and Gregory drew the people's attention to 'the glowing light and dwindling darkness that follow the Nativity', while Leo the Great denounced the idea that Christmas Day is to be honoured not for the birth of Christ but for the rising of the sun. But the old paganism died hard and the Christian teachers found it unwise to attempt a complete break with cherished practices. Sun-worship lies at the heart of great Christian festivals, the heathen relics blending with beliefs antagonistic to the spirit that prompted them, and nothing proves more clearly the appeal of solar idolatry than the efforts made by Moses to prohibit it. 'Take care,' he cried to the Israelites, 'lest when you lift your eyes to Heaven and see the sun, the moon, and all the stars, you be seduced!' When today we light our candles on our Christmas trees it is but a remnant of the rite which was to guide the sun-god back to life; and when we eat our plum-pudding we realize that though it is an act grossly incongruous with the sorrow and suffering of Jesus, it is in harmony with a feast in honour of the sun as held in ancient times when cakes of corn and fruit were laid on altars dedicated to the Lord of Light. And as a matter of fact most Christian peoples to this day prefer to celebrate Christmas frankly in the manner of a pagan festival.

Since the sun rises in the East, that came to be considered the most propitious direction to face on ritualistic occasions, and the word Easter is as clearly related to sun-movements as the word East, while in China there was a popular belief that if the Emperor would steadfastly turn his countenance towards the East there would be perfect harmony on earth. In Europe we face the East when reciting the Creed, in Mexico when kneeling in prayer, and a certain Baptismal rite ensured that he who embraced the Faith first turned towards the West and renounced satan with gestures of abhorrence. We know how the architects of St Peter's in Rome so exactly placed the Basilica due East and West that on the vernal equinox the great doors could be thrown open at sunrise and the rays thus penetrate through the nave and illuminate the High Altar, just as in pagan days the granite blocks were so arranged at Stonehenge that at the summer solstice the shadow of one stone fell exactly on the stone in the centre of the circle, indicating to the priests that the new year had begun and the signal could be given for flashing the news through the land.

We find solar significance in words and actions unrecognized as such by the conscious mind. Thus the most ancient symbol of the sun is a wheel with spokes for rays, and if we dig into the derivation we find that our word wheel comes from Yole or Yuul, Hiaul or Huul, meaning Sun and used as a festival term at the winter solstice, so that we still speak of the Yule Log. We find relics of sun-worship in many of our most familiar signs. It was the custom of painters when representing the head of Christ, or the Virgin, or even that of some particularly saint-like person, to place upon it the solar circle or disc, while we still speak of a man as 'being worthy of a halo'. The simple cross with perpendicular and transverse arms of equal length representing the spoke of the solar wheel sending out rays in all directions found upon the monuments and utensils of every primitive people, evolved into the symbol of the Cross which because of its sadness and poignancy held an immense appeal for suffering mankind, and came at last to stand for the price of redemption through the Christian scheme of salvation; and we all know well enough how in 1939, on account of the Christian tendency towards forgiveness, meekness, and appeasement, that other emblem of the vernal sun, the Swastika, signifying vitality, health, violence, ruthlessness, power, and glory, was raised against the Cross in mortal combat by Hitler's heathen hordes.

4. The Scientific Approach

When we think of our modern sun-temples we might well bear in mind the Egyptian solar religion. Astonished by the sublime order of the universe they exalted the sun as the central object in significance, naming it Osiris, also called Ra, with its eternal and primary companion Isis, the moon, and set up great temples such as that of Amen-Ra, at Karnak, 'One of the most soul-inspiring temples which have ever been conceived or built by man,' says Sir Norman Lockyer. Covering twice the area of St Peter's, its vastness is unchallenged in the ecclesiastical world; straight and true its stone avenue runs for five hundred yards, thus limiting the light into one narrow beam carried to the other extremity of the building into the sanctuary to illuminate at the solstice the image of Ra.

Our sun-temple is an observatory. Our belief, far outstripping our belief in Scripture, is in Science. We are inclined to think that the Egyptian shrine only expressed ignorance of the true nature of the sun which is not a god but a ball of fire ninety-three million miles off. Yet we might fairly put it another way. They were under an illusion with regard to the natural facts about the sun, but their Knowledge, which with a capital K may be taken as implying awareness of the total significance concerning phenomena, was not inferior to ours. We know the facts. We know a tremendous amount of facts. They take the place of significance in our minds. Think of that most modern of all productions, the *Encyclopaedia Britannica*. Think of its weight upon the intellect, and of its deliberate, its planned refusal to contain a single sentence set down with the purpose of exciting total awareness, as if that part of man's mind, the imagination, is worth nothing. Think of the encouragement given nowadays to millions through crossword and quiz to suppose that their bits and pieces of information constitute knowledge; and imagine the consternation and scandal on the faces of quizmaster and quizzee, if, on a single occasion it were suggested that the prize should be handed to the man who had smartly answered questions on mythology, only on condition that he could give an intelligent answer as to the significance and meaning of myth!

Is there much point in classifying everything and being aware of nothing? Yes; on that basis we can do a great deal. Once in possession of this factual knowledge about phenomena we can build up our civilization with its advantages and comforts. That is something – even if we endanger it all with that bomb we dislike so much and make so eagerly. But we cannot expect to have much in the way of religion (apart from theologies). And of course we can be just as credulous as our ancestors, in the modern manner, ready to take as gospel the scriptures of science, and to accept without batting an eyelid, for instance, the announcement that the solar system is merely one 'island' amongst millions in space travelling from nowhere to nowhere at a thousand miles a second. There is little to choose between on the one hand the Aristotelians who insisted that the findings of Galileo, Scheiner, and Fabricius concerning the spots on the sun were a delusion and that it was out of keeping with the dignity of the Eye of the Universe to suppose that it could be inflicted with so plebeian a complaint as opthalmia, and on the other hand the Rev. Tobias Snowden who commanded a considerable following in the nineteenth century when he published a book proving the sun to be Hell and the dark spots gatherings of damned souls; while we ourselves swallow with little demur the ruling of the experts who relate the periodicity of sun-spots and turbulence with affairs on earth such as the migration of swallows, the yield of harvests, and the revolutions of states. Unfortunately our modern credulity is not accompanied by any sense of the *numinous*, of what used to be called 'participation', of reverence, of that 'experience of awe before the pure phenomena' which Goethe considered to be the highest faculty in man.

There is nothing abstruse or even fresh in these remarks. We are accustomed to hear them made. 'A lot of people think that a little peasant boy of the present day who goes to a primary school knows more than Pythagoras did, simply because he can repeat, parrot-wise, that the earth moves round the sun,' said Simone Weil in *The Need for Roots*, and adds: 'In actual fact he no longer looks up at the heavens. The sun about which they talk to him in class hasn't, for him, the slightest connection with the one he can see. He is severed from the universe surrounding him.' But I cannot agree that such serverance is aprerequisite of modern knowledge. There is no inevitable dichotomy between the scientific and the religious approach – for myself, I feel no such 'split'. There is no reason why the facts which great and devoted men lay before us should be regarded as alien to the soul, or as 'mere facts' useful only for practical purposes or for passing examinations. They can equally well feed the imagination. The split is not real; but to suppose its existence has been easier since the coming of Newton.

CHAPTER III

SCIENCE AND IMAGINATION

1. Newton and the Poets

On 28 December 1817 Wordsworth, Keats, Lamb, and Benjamin Haydon dined together. They discussed the merits of Homer, Shakespeare, Milton, and Virgil. As the evening advanced Lamb became slightly the better for drink, and as a set off to the solemnity of Wordsworth he abused Haydon for putting Newton's head into a picture he had just finished – 'a fellow who believed nothing unless it was as clear as three sides of a triangle.' Then he and Keats agreed that Newton had destroyed all the poetry of the rainbow by reducing it to its prismatic colours. They all ended with a toast, 'drinking Newton's health and confusion to mathematics'.[1]

I'm not quite sure why they drank Newton's health under the circumstances, but let that pass. Though in their cups – and perhaps maudlin – their tone was true to the general feeling that science was no longer on the side of poetry, nay, that it was the enemy of creative imagination. Today no one would be surprised at such an attitude, but it was not always so. In the Middle Ages the 'sciences' were 'the arts of the mind'. Every branch of learning was a *scientia*, and whether the subject was grammar, logic, geometry, music or other things, the term 'sciences' or 'arts' could be used. We still pay letter-tribute to the sense of this. I myself am a Bachelor of Arts. When I got it I was considered as having achieved a degree – of what? In what lay this degree of excellence? I could not have claimed it in Art, still less in Science. But the implications are sound and pleasant, however unworthy the recipient, and remind us of the old days when 'science was used to define knowledge generally, the state or fact of knowing', as Sir Ifor Evans puts it, 'while Philosophy

1. *The Autobiography and Memoirs of Benjamin Haydon.*

was still widely used to define Science as we now understand it'.[2] In fact 'a philosophical apparatus' meant a scientific apparatus.

When in the sixteenth century Sir Philip Sidney spoke of a tale which holdeth children from play and old men from the chimney corner, it was not of stories he was thinking, but of knowledge. In the seventeenth century there was little or no conflict between science and literature. Science nourished the works of Sir Thomas Browne; Robert Hooke's *Micrographia* is a real work of art; and perhaps my brief quotations in the previous chapter from Baker and Wilkins and Adams and even Leeuwenhoeck will suggest that even those who had no pretensions as literary men yet wrote with considerable felicity under the inspiration of the simple science of the day. Richard Baxter, Jeremy Taylor, and Joseph Glanville all approached science with pleasure and without misgiving. It was possible for a man such as Samuel Pepys to be President of the Royal Society; Dryden, Waller, and Denham were happy to associate themselves with it; and Abraham Cowley, who was among the first to be nominated for the Society, had great influence and significance in his day.

Newton was a seventeenth-century, early eighteenth-century man (1642–1727). His *Principia* was written in Latin and was in any case beyond the understanding of literary men, but his *Optics* was understood, and at first it atttracted the writers of the day. In fact the minor poets composed pieces in his praise, the best known being James Thomson's 'Ode to the Memory of Sir Isaac Newton',[3] elaborately eulogizing his theory of light. Newton gave Pope material for poetry and Swift material for satire. Pope was so great an artist that almost anything he had to say reads as well now as two hundred years ago. Indeed some of his lines gather force year by year, perhaps the most pertinent example appearing in Book II in the *Essay on Man:*

> Go wondrous creature! mount where Science guides
> Go, measure earth, weigh air, and state the tides,

2. *Literature and Science.*
3. Marjorie Nicolson: *Newton Demands the Muse.*

> Instruct the planets in what orbs to run,
> Correct old Time, and regulate the sun;
> Go teach Eternal Wisdom how to rule –
> Then drop into thyself, and be a fool.

This is criticism of man and not a reaction against science. The point and strength of the *Essay* lies in the fact that the muse, so far from being weakened by science, acquires material to work on, even if in terms of criticism. The same might be said of Donne over a century earlier. His famous lines:

> And new Philosophy calls all in doubt
> The Element of fire is quite put out;
> The Sun is lost, and th' earth and no man's wit
> Can well direct him where to look for it . . .

are often quoted as being written by a man who felt that the new learning introduced with Galileo had thrown men into confusion. But it may be questioned whether Donne was really very perturbed. He was an exceedingly clever man, and I suspect he enjoyed the opportunity for epigram and was glad to use science to fertilize his rhyme. It was the same with Swift (1667–1745). It was all grist to his mill. Being no respecter of persons he did not respect Newton (as Pope did) nor trouble to understand him. He simply had his fun, especially in 'The Voyage to Laputa' in *Gulliver's Travels*. Pope wears much better than Swift. In spite of his celebrated prose-style Swift is, for the most part, tedious now: he seems off the point as far as we are concerned.

2. Blake and Wordsworth

After Newton had been dead for thirty years William Blake was born (1757–1827) who declared war upon him for twenty years; after he had been dead for twenty-two years Goethe was born (1749–1832) who declared war upon him for thirty-two years. The two poets have nothing in common save this enmity to

Newton – and I must let Goethe stand down for the moment. Blake was the first great poet to propose manifestos against science; he was the first to get hold of a knife and cut people's heads up. He made a tremendous slash through the skull neatly dividing the cranium into two equal parts. In one half he discovered vision, imagination, intuition; in the other half he discovered reason, analysis, and the desire for experiment: one was spiritual, the other materialist; one was right, the other wrong; one was good, the other bad. The way of analysis was the way of corruption and mockery, the words 'reason' and 'experiment' were the slogans of the devil; and this was the way of Descartes and of Locke, and these were the slogans of Bacon and Newton.

> Mock on, Mock on Voltaire, Rousseau:
> Mock on, Mock on: 'tis all in vain!
> You throw the sand against the wind,
> And the wind blows it back again.
>
> And every sand becomes a Gem
> Reflected in the beams divine;
> Blown back they blind the mocking Eye,
> But still in Israel's path they shine.
>
> The Atoms of Democritus
> And Newton's particles of Light
> Are sands upon the Red sea shore,
> Where Israel's tents do shine so bright.

The poem is slightly obscure (in the right way of being obscure) but we would not have it otherwise. Blake's particular kind of inspiration, his God-intoxication, his immediacy of response could not be nourished by science and might very well be injured by analysis so that a grain of sand would cease to be a gem reflected in the beams divine. When he looked at the setting sun he didn't see 'a little round disc about the size of a shilling. Oh no, I see a company of the Heavenly Host crying – Holy, holy, holy Lord God Almighty!' We do not take him literally but it is clear that his inspiration was that order which could not be supported by the scientific approach. It is also clear that he was upset by the theory of primary and secondary qualities elaborated by Descartes and Locke. Blake was one of the most glorious poets who ever existed, and I am unwilling to bother about his polemics. He was easily enraged. Even Wordsworth's pantheism gave him a severe stomach complaint.

The position of Wordsworth in this matter is ironical. It is true that his famous lines on Newton 'with his prism and silent face, The marble index of a mind for ever Voyaging through strange seas of Thought, alone' are never forgotten (who could forget them?) but it seems that he is firmly fixed as having declared that it was murder to dissect, and that a scientist is the man who would peep and botanize upon his mother's grave; though if we examine the context of the latter remark we find that he was simply making a plea for regarding individuals as individuals and not as types in the sociological Sidney Webb manner, while surely any poet, conscious of his intense experience of beauty, may be allowed to mutter the words, 'We murder to dissect,' faced with moralists and analysts pure and simple. Less well known are the following words from his introduction to *The Lyrical Ballads:* he looks to the time when 'the remotest discoveries of the Chemist, the Botanist, or Mineralogist will be as proper objects of the Poet's art as any upon which it can be employed. If the time should ever come when what is now called Science, thus familiarized to men shall be ready to put on, as it were, a form of flesh and blood, the poet will lend his divine spirit to aid the transfiguration.' Those words might well be used today as the apologia or manifesto of those who believe that the time has now come for the poet to exhibit before the public gaze the spoils and trophies of the investigators.

But that does not complete the irony of Wordsworth's position. It happens that he is the poetic pet lamb chosen by scientists who are anxious to show how the poets support their theories. The best example of this may be found in Whitehead's chapter on 'The Romantic Reaction' in his *Science and the*

Modern World. He quotes one of Wordsworth's famous apostrophes to Nature in *The Prelude* with approval as 'exhibiting entwined prehensive unities, each suffused with model presences of others.' Indeed it may be so. But had that been the main point of the poem Wordsworth might just as well have written in prose. The point lay in the experience of joy which this consciousness brought with it. It may please Whitehead that Wordsworth lent support to his theory of prehensive unification and even of misplaced concreteness, but neither Wordsworth nor any other poet minds very much whether or not his exalted experience is supported by the theories of science. I do not deny the intellectual satisfaction which this synthesis undoubtedly affords. It gives pleasure to many people. I often hear people say: 'Things are looking up! The scientists have abandoned their old theories of matter and are now declaring that it is vanishing before their eyes.' They are pleased when they say this. Their eyes sparkle. They smile. But what are they actually saying? Only that what was hard is now soft. That what seemed to be a solid slab is really a whirling mass of electric particles. They think that this makes matter less material and more spiritual. They equate electricity with spirit. Thus the more electricity we have the more spirit we have. I am not sure where this leads us. If it is true it will be of little concern whether we call ourselves materialists or idealists. Of course it may be true. Many good men have derived satisfaction from this thought. *'Spirit is matter seen in a stronger light'*, suggests L. P. Jacks with emphasis, and adds that when the late Sir William Bragg, President of the Royal Society, maintained that 'light is the basis of matter and the substance of the Whole', he was giving 'another wording for Spinoza's Substance'.[4] It is more than likely. All I am saying is that we are at an advantage if we experience these things – as did Spinoza – rather than only reason about their possibility. A drop of truly experienced harmony is worth an ocean of intellectual synthesis. Experience is a bigger gun than theory in this sphere, and just as the man with strong religious intimations is not in the least parasitic upon the

4. 'They Do Ill Who Leave This Out' in *Near the Brink*.

validity of any historic event, so the poet in his exultation is independent of any scientific theory. Dependence here upon the changing scientific approaches leads to confusion. It is deplorable to see learned men of strong minds – if they really are strong – plucking metaphysical comfort from the Quantum Theory on the ground that it suggests that Nature is rather less a determined system than we had thought – a conclusion which seems to me as absurd as the comfort. I am relieved to find that I am not singular in maintaining this view. 'While I think that the argument which finds evidence for human free-will in the supposed discovery of indeterminacy in Nature is entirely groundless,' writes Professor Herbert Dingle, 'I think that the nineteenth-century argument against human free-will because of the supposed determinacy in Nature is equally groundless. Our consciousness of freedom is an immediate experience, and as such is a fundamental datum. It is inconceivable that experience can be refuted by deductions from experience.'[5] That observation might fruitfully be contrasted with the long speech in Shaw's *Too True To Be Good* delivered by the distracted Elder through whom the great playwright so faithfully mirrored the confusion of our time.

3. Keats and the Rainbow

This brings me back to Keats and the rainbow. It may seem a little strange that the man who claimed that the experience of Beauty obliterated all other considerations, and was in fact the answer to those questions which we bundle under the head of Truth, should have been in the slightest degree disturbed by the discovery of the dispersion of light as exhibited by the rainbow. It would be absurd to quote seriously a remark dropped between drinks at a dinner party, such as the idea that 'Newton had destroyed all the poetry of the rainbow by reducing it to its prismatic colours'; but Keats did elaborate the charge in *Lamia II* with the lines:

5. 'The New Outlook in Physics' in *The Scientific Adventure*.

> Do not all charms fly
> At the mere touch of cold philosophy?
> There was an awful rainbow once in heaven:
> We know her woof, her texture; she is given
> In the dull catalogue of common things.
> Philosophy will clip an Angel's wings,
> Conquer all mysteries by rule and line,
> Empty the haunted air and gnomed mine –
> Unweave a rainbow . . .

That is a real 'murder to dissect' attitude. In passing, we may wonder what Shelley thought of those lines. For Shelley loved science and was greatly influenced by astronomy and attracted to chemistry. 'It symbolizes to him joy, and peace, and illumination,' says Whitehead. 'What the hills were to the youth of Wordsworth, a chemical laboratory was to Shelley. It is unfortunate that Shelley's literary critics have, in this respect, so little of Shelley in their own mentality. They tend to treat as a casual oddity of Shelley's nature what was, in fact, part of the main structure of his mind, permeating his poetry through and through. If Shelley had been born a hundred years later, the twentieth century would have seen a Newton amongst chemists.' That is a good gambit: and we are bound to admire that 'would have'. He goes on to illustrate the permeation of Shelley's poetry by science by giving as example 'the vaporous exultation not to be confined' as the exact poetic transcript of 'the expansive force of gases'; followed by eight lines of extraordinary support for six lines of extraordinarily weak verse. His claim is somewhat dubious, just as it is dubious to suggest, as others have, that Shelley's 'The Cloud' could not have been written without a knowledge of meteorology, or his lines, 'Life like a dome of many-coloured glass, Stains the white radiance of Eternity,' without benefit of Newton's prism. However, Whitehead's main point is that Shelley who was so sympathetic to science and absorbed in its ideas was not in the smallest degree influenced by the doctrine of secondary qualities which had made Blake so angry. He simply ignored it.

I do not think that Keats was distressed by what Locke or

Descartes had to say, he may not have read them. But he does appear to have been bothered by Newton's explanation of colours and hence of the rainbow. He has unwoven it, Keats says; he has taken it to pieces; he has dissected it. Therefore he has murdered it. Its beauty is gone. It is now in the dull catalogue of common things. But why? (And is any common thing really dull?) This is an odd point of view for a poet. I take it that my own point of view is the more normal: I rejoice at the red of the rose. Then I am severely told that nothing is red. I am told that no colour is 'a fact in external nature'. I am told that redness is merely motion of material; that the light striking my eyes is merely motion of material; that sound striking my ears is merely motion of material; that scent striking my nostrils is merely motion of material. I am told that without eyes, ears, or noses there would be no colour, sound, or scent in the universe. I am supposed to be depressed. But I am easily pleased. I had thought that the redness belonged to the rose – and I was glad. I am now told that it was eight minutes ago in the sun – and again I am glad, since this strikes me as being even more miraculous. Colour may not be strictly a 'fact'. It is allowed to be an 'event'. It remains a reality: and not less a reality because my own mind helps to shape it. The rainbow is still a rainbow 'considered by the theologian to be a wonderful sign of God and by the scientist to be a sign of the Laws governing God's creations', as Spinoza said.[6] When we are told new things about Nature our sense of wonder is increased not decreased. When we learn about colours we do not thereby think that the holy chalice of the yellow tulip is less pure a cup; that the candelabra of the chestnut trees burn with less bright a flame; that the peacock is robbed of its prodigy of plumage; nor fail to see Stevenson's 'ragged moor receive The incomparable pomp of eve'.

6. *Reeckering van den Regenboog* in *Opera Quotquot Reperta Sunt*.

4. Science and Imagination

I think we are inclined to become enslaved by categories. We imagine we are imprisoned behind barriers that do not necessarily exist. Thus we speak of the scientific approach over against the poetic approach to reality, and of the analytical mind over against the creative mind, and of reason against intuition. It is true that these faculties are realities. It is true that there is the knowledge which comes with ecstasy and the knowledge which comes with rational effort. There is the way of analysis and there is the way of imagination. But one lot does not exist in this kind of person and the other lot in that kind of person. Most of us use both and they are not in neat packets in our heads either. Consider. I sit in the garden under a pear tree. I see an animal in front of me in the air. It is not a bird, it has no wings, it cannot fly; but it is getting on extremely well, and has comfortably taken a seat upon nothing. By way of experiment I give it a little prod, and at once it rises as if in a lift to a twig above my head. Presently it comes down in the lift again. As I gaze at this creature we call a spider I see that it puts up as good a show as anything to be seen on earth. Now what has been my attitude, my approach? Have I been scientific or poetical? Have I bestowed the analytical or imaginative faculty upon it? I don't know. Quite honestly, I don't mind. The probability is that I have used an amalgam of all the faculties I possess in my endeavour to really see the little creature.

I do not think we can put the true scientist on the left and the true poet on the right and think of them as in different categories. They are in the same category: they are both artists. In both the main faculty is imagination. Imagination is the power to see what is there. Once an agricultural labourer said to me after an aeroplane had crashed in an adjoining field: 'It wouldn't do for anyone as has too much imagination to go over and look at it, for he might be seeing things as aren't there.' But that is what imagination is *not*. It is the power to see what is there. Fancy, invention, is the capacity to see what is not there.

Imagination sees what is there with full concentration in combination with the faculty of love – indeed imagination has also been defined as 'Intellectual Love' and as 'Reason in her most exalted mood'. This is man's highest faculty, the power to see, to be a seer, to fasten upon the total significance of phenomena – and even to image further. It was by his 'wonderful imagination', we are told, that Newton was constantly discerning new tracks and new processes in the region of the unknown. This imaging into the centre of reality seems to belong to the great scientists as much as to the great poets. There is nothing to choose here in force of imaginative power between a Rutherford who can penetrate into the very heart of matter and a Tolstoy who can penetrate in to the very heart of man.

The distinction between science and art is 'purely conventional' said Herbert Spenser, 'for it is impossible to say where art ends and science begins'.[7] The learned historian of science, Charles Singer, defined it as 'knowledge in the making' and as 'the growing edge between the unknown and the known'. That is to say it is active and creative, essentially artistic. We are not talking about the man who applies science over against the creative artist, but of two sorts of artist. It has been claimed, surely with truth, that there is no such thing as 'an unimaginative scientific man'. John Tyndall said of Faraday that 'the force of his imagination was enormous', and it has always seemed to me that Tyndall himself was a good example on a lower plane; he says how he could hardly sustain with any degree of calmness the beauty of phenomena revealed by his physical researches. 'The true spirit of delight,' said Bertrand Russell, 'the exaltation, the sense of being more than man, which is the touchstone of the highest excellence, is to be found in mathematics as surely as in poetry' – more so perhaps, judging from some modern poets.

Sometimes it is far from certain whether we should call a given man a scientist or an artist. Ruskin claimed to possess the most analytical mind of his time. He is thought of as an art-critic, I believe. He was primarily a geologist who liked paint-

7. His essay, 'The Genesis of Science'.

ing. His own drawings seem to me wonderful, and at times he was a great poet in prose: in few men have science and literature been so integrated, however undisciplined and unwhole his individual books. When we think of Leonardo da Vinci we do not know what to call him. Indeed his case is still held over. It is not yet decided whether he should be called the supreme painter of his day or the supreme scientist. Some have suggested that it is almost as plausible to regard Leonardo as primarily an engineer as primarily a painter. 'The dispute as to whether he was above all an artist or a man of science is a foolish or even an unmeaning dispute,' wrote Havelock Ellis, himself an example of the artist integrated with the scientific investigator and philosopher. He thinks that Leonardo's painting was only a concession to his age, and that from youth to old age he had directed his whole strength to one end: the knowledge and mastery of Nature. The medium in which he worked was Nature, the medium in which the scientist works; 'every problem in painting was to Leonardo a problem in science, every problem in physics he approached in the spirit of the artist.' Finally, Havelock Ellis compares him with Newton:

He seemed to himself to be, here and always, a man standing at the mouth of a gloomy cavern of Nature with arched back, one hand resting on his knee and the other shading his eyes, as he peers intently into the darkness, possessed by fear and desire, fear of the threatening gloom of the cavern, desire to discover what miracle it might hold. We are far here from the traditional attitude of the painter; we are nearer to the attitude of that great seeker into the mysteries of Nature, one of the very few born of women to whom we can ever passingly compare Leonardo, who felt in old age that he had only been a child gathering shells and pebbles on the shore of the great ocean of truth.[8]

It is true that Newton did not appreciate art in the restricted sense and declared that sculpture was no better than the making of stone dolls. It is true that Darwin in his *Autobiography* lamented his increasing incapacity to endure a line of poetry, and said that when he tried to read Shakespeare he found him so

8. *The Dance of Life.*

intolerably dull that he was nauseated. It is sad, of course, that Newton and Darwin were so narrow in their approach, but this has little to do with our present proposals. 'Darwin was one of those elect persons,' to quote Ellis again, 'in whose subconscious, if not their conscious, nature has implanted the realization that science *is* poetry, and in a field altogether remote from the poetry and art of convention he was alike poet and artist.' Such a remark could be easily misinterpreted and still more easily misapplied, but personally I would accept it heartily, for I really do think that it is maddening nonsense to suppose that the shaper and creator of thought is less of an artist than the creator of design. The trouble arises owing to the existence of the bogus gentlemen. There are many who claim the title of scientist without any right to it – men who are mere compilers of facts. These people are no more loved by genuine scientists than the mere bookmakers cluttering up the literary scene (men who in happier days, before ignorance and iliteracy had made such rapid strides, would have been peacefully employed in copying illuminated MSS) are loved by genuine literary men. These 'scientists' are always the first to decry and even to prove the 'impossibility' of imaginative proposals such as the introduction of the telegraph, or the telephone, or the airship, or the steam locomotive, and so on. These are the people who do actually approach reality in a completely analytical and abstracting manner which spells death to the quality and value of the thing-in-itself, so that, for example, as Eddington pointed out, an elephant sliding down a grassy hillside on to a bed of softly yielding turf becomes 'a mass of two tons' sinking 'at an angle of 60 degrees' on to 'a coefficient of friction'.[9] Pointer-readings take the place of the reality (or poetry) of the object just as to a musketry instructor that tree on the hill is simply 'a definite object' to be fired at. These are the people who, habitually exercising their minds in this manner, do really come to believe that knowledge is a question of weighing, measuring, and counting. They even become unaware that the object's property of quality is its chief truth for humanity at large. If

9. Eddington: *The Nature of the Physical World*, pp. 251–3.

you argue with them they tell you that you are 'emotionally involved' as if that proved something against you. Their pointer-readings are skilful and useful, but the habitual attitude of mind engendered weakens their intelligence. It is low grade when off the speciality – not much above that of a garage assistant. When they write they do so without personal touch or quality, so that if one has to review a symposium of their essays, there is no distinguishing one from another. Actually, they despise language. They do really think that instead of saying 'Let there be light', it would be better to say, 'Let the molecules swing together effectively'. They feel that that is being tough and hard-boiled. It is indeed: but it is the toughness of over-cooked beef; it is the hard-boiledness of an overdone egg – in both cases indigestible.

The truth is that from early youth they have been forced to concentrate upon the part. In the end they are never able to see the whole. A man has a better chance if he first attains a vision of the whole and then comes to the part. Still, I must add that, in my experience, many young scientists are not content with their partiality and show an interest in the humanities. They compare favourably with the literary men who take pride in asserting that they 'don't know a thing about science'.

5. Goethe's Attack on Newton

At this point the reader may feel inclined to ask how, if there is no conflict between the true scientist and the true poet, it came about that Goethe carried on a feud against Newton for thirty years? Since Goethe is perhaps the most celebrated name in literature for combining poetic sensibility with an absorbed interest in science, this conflict may seem a little odd. It might be thought that the child who was thrilled with Franklin's doctrine of electricity and that the man who, among other scientific interests and experiments, proposed a botanical theory of the metamorphoses of plants, would have been at ease with Newton. Yet it is not so strange, for the fact is that he was too much at ease and felt confident that he knew better. He did have something of the greatest importance to say. His method of doing so, however, was curious, and was not free from some slight element of comedy.

From early youth Goethe had enjoyed theorizing on painting, though he could not paint (it was unnecessary for him to theorize on poetry). So he came to colour. What is colour? He was referred to Newton. One day Professor Büttner lent him some prisms and optical instruments to carry out the prescribed experiments. He put off using them. It is rather strange, this. The great poet and science student for months was too slack even to take a prism in his hand. He was repeatedly asked to return the equipment, and at last received a definite demand that they must be returned at once. So he roused himself, took one prism and looked through it at the white wall of his room. He expected to see, and according to the rules he thought he should see a rainbow-coloured wall. But he did not. The wall remained white – with coloured edges. The experiment had failed to work for him. Immediately he exclaimed aloud: *'Newton's theory is false!'*

It is rather interesting. Newton, the acknowledged supreme intellect of the seventeenth century, after endless labour had interpreted the heavens and examined the constitution of light, and for nearly a hundred years his theory had stood the test. Then Goethe glances through a prism for a few minutes, fails to see what he expects, and instantly declares that Newton is wrong. The plain truth is Goethe had all the arrogance of the man of position and reputation. We are reminded of Tolstoy; or rather, of Tolstoy in his old age as seen by Chekhov. 'All great sages', said the latter, 'are as despotic as generals, and as ignorant and indelicate as generals, because they are confident of impunity. Diogenes spat in people's beards fully aware that he would not be punished for it. Tolstoy abuses doctors as scoundrels and allows himself to remain in ignorance of great questions because he is just such a Diogenes who won't be taken to the police-court nor be abused in the papers.'[10] So with

10. Chekhov: *Letters.*

Goethe. Newton had got it all wrong. He had perpetrated an experimental incoherence. To try to analyse light was a shallow blunder. To suppose that white light could contain coloured light or that when coloured rays were squashed together they would look white, was to suppose an absurdity, to indulge in fairy-tales. To pass a poverty-stricken thread of light through a tiny hole into a dark room, when by going out into the open air any amount of it could be had free of charge, was ridiculous. He elaborated his own *Theory of Colours*. Colours originate, he said, in the modification of light by outward circumstances. They are not developed *out* of light but *by* light. For the phenomenon of colour there is required darkness as well as light. (Personally, I can get the colours without any darkness in the bright spray giving me my rainbow in the garden.) That was the gist of his Theory. He never looked back. His was 'the pure doctrine'. His opponents were simply people who 'continued in error'. All that was necessary to understand it was 'a sound head' – in contradistinction to the imperfect intelligence of Newton. He repeatedly told Eckermann that he took no pride for what he had done as a poet, but that in his century he was the only person who knew the truth in the difficult science of colours – of that he was not a little proud. He had come into a great inheritance: Napoleon had inherited the French Revolution; Luther had inherited the darkness of the Popes; and he had inherited the errors of Newtonian theory.

Now I have no intention whatever of spoiling my book at this stage by being drawn into this controversy on the physics of light. In any case I do not possess the equipment. It also seems clear that Goethe himself did not know much more about physics, *per se*, than he did about ornithology (remember his endearing remark to Eckermann that he could not distinguish a lark from a finch). No layman can move freely in the field of light-dispersion without detailed study – let alone propose a theory of his own. When I first acquainted myself with this business of Goethe and the prism I thought it only sensible to conduct the same experiment myself and find out what I personally saw on looking through a prism at a white wall. I did so in my own room; and I saw – *exactly what Goethe saw*: a white wall with coloured edges. I was neither pleased nor sorry. But I went a step further than Goethe. I stepped into the garden. I hung a white towel on the laundry line. Standing three yards away I looked at it through the prism. Again I saw what I had seen in the room. Then I retreated thirty yards distance from the towel and looked at it once more. This time it was *fully rainbowed* except for a little white in the middle.

Even if I could I would not, here, go into the reasons for this – reasons which are explained by Newton in terms of 'pure and impure' spectra. For I do not mind about it. If I were obliged to teach anyone anything or pass any examination, I would mind – otherwise I prefer to let it go. You may think that my refusal to worry further in this matter exhibits a deplorably unscientific frame of mind. Yet, speaking quite personally – but hoping that I am speaking for some others as well – I have now enough here to nourish the imagination. That is the motive of my studies. Carlyle offended Ruskin by going to North Wales for two months 'and noting absolutely no Cambrian thing or event, but only increase of Carlylian bile'.[11] I am not as bad as that. Carlyle also pained Darwin by saying that 'he thought it a most ridiculous thing that anyone should care whether a glacier moved a little quicker or a little slower or moved at all'.[12] In fact he could not care less. I could not agree more – *au fond*. For though I happen to have studied the glacier question carefully and the controversies about it, I really do not care a bit about the rate of movement. Science helps us to see more in a given thing than we would otherwise see; it puts us through the discipline of factual concentration, which is the best possible way to feed and enrich the imagination. In my own case when it has once done this I am satisfied and can dispense with further facts and controversies. So with the theories of colour: I confess that I am supremely indifferent as to whether colours arrive *in* the light or *by* the light; it is the fact that they have arrived at all that holds me; it is in the finished article that I am finally

11. Letter to E.N.C.
12. Darwin's *Autobiography*.

interested. Perhaps some of my readers may agree with me, and they may wonder why Goethe was so concerned with the mechanism of the business; they may wonder what philosophical or religious significance the actual mechanism could possibly have for him or anyone else.

That brings me to my main point, and my reason for introducing this controversy at all. For Goethe feared that the scientific descriptions of mechanism would promote a philosophy of mechanical materialism. He put forward his views without the force of economy, without wit, without grace, and without clarity – yet he has found support from that day to this. He was hailed by Hegel, by Schelling, and later by Schopenhauer who wrote an elaborate defence of his theory,[13] while in our own day he has received the massive support of the great Rudolf Steiner who attacked the new physics in a manner fully as puzzling but much more amusing: 'The theory of the colour of solids,' he writes, 'is worthy of the new physics. Why is a body red? A body is red because it absorbs all other colours and reflects only red. This is the explanation so characteristic of the new physics, for it is based approximately on the logical formula: Why is a man stupid? He is stupid because he absorbs all cleverness and radiates only stupidity outwards. If one applies this logical principle so common in colour-theory everywhere to the rest of life, you see what interesting things result.'[14] But he does not say why anyone else should wish to apply this principle to the rest of life, or why it is wrong to apply it to colour.

It may give some of us on this side of the Channel pleasure to turn for light to our own dear Coleridge who was kind enough to separate the threads. 'If it please the Almighty to grant me health, hope, and a steady mind,' he wrote to Poole on 23 March 1801, 'before my thirtieth year I will thoroughly understand the whole of Newton's works. At present I must content myself with endeavouring to make myself entire master of his easier work, that on Optics. I am exceedingly delighted with the beauty and neatness of his experiments, and with the accuracy

13. *Van der Farben in Schriften zur Erkenntnislehre* (1873).
14. *Colour.*

of his *immediate* deductions from them.' That is a pleasing approach to Newton, and we are the more ready to attend to his criticism. He goes on:

But the opinions founded on these deductions, and indeed his whole theory is, I am persuaded, so exceedingly superficial as without impropriety to be deemed false. Newton was a mere materialist. *Mind*, in his system, is always passive – to a lazy *Looker-on* on an external world. If the mind be not *passive*, if it be indeed made in God's Image, and that, too, in the sublimest sense, the *Image of the Creator,* there is ground for suspicion that any system built on the passiveness of the mind must be false, as a system.[15]

We can all rejoice in the beautiful mechanism of optics. The whole vegetable world has no eyes, can see nothing of its own beauty. Some animals have no sight: to this day the worm, for instance, is without eyes and would not know what to do with a pair. However, in the course of evolution the bodies of most animals became sensitive to radiation and the molecules responded in a special manner to light-waves. As time went on and larger animals developed they lost the sensitivity to this force over the whole body and concentrated it in a particular place – a little patch sensitive to light. This patch grew into a sense organ entirely responding to light. After countless ages it evolved into the Eye – serving alike the girl who also speaks (or listens) with her eyes and for the rattlesnake's deadly cleft in the glazed blue of the ghastly lens; for eyes perched on pyramids of bone or waving on the vulnerable points of the snail's pillars; for eyes brandished on horns or massed in clusters.

We can go inside an eye and have a look round just as if we were entering a room. In fact the way to do this is by entering a room. We go in, shut the door, and pull down the blinds. Opposite the window we put up a white screen. Rude as this equipment is, it is all we need in order to sit inside an eye. We need only do one other thing – make a little hole in the middle of the blind, and take a seat in a corner of the room. Then immediately, because light is not something stationary, because it is the

15. Letter XLVI in the Nonesuch *Coleridge*.

arrival, the continuous arrival of energy, it forces its way through the little hole in the blind. If outside the window there are a tree, a cottage, a pond, a field, and a man, all scattering the rays of the sun, then from every part of those objects rays will pass and print themselves upon the screen – and we have a *picture* of the scene outside. We are sitting inside a camera, however crude. We are sitting inside an eye, however rudimentary. For room read dark walls of camera; for screen read photographic plate; for blind read shutter; for hole read lens. And then for walls of camera read walls of eye; for photographic plate read retina; for shutter read iris; for lens read eye.

(In parenthesis: it seems strange at first that instead of the little round hole in the blind being printed on the screen, the whole scene outside should manage to make its way through the opening. Luckily we need not take this secondhand: we can prove it for ourselves without any laboratory. Thus I take the tray out of an empty match-box, stick some white paper across the open side, pierce a hole with a needle through the middle of the bottom of the tray – opposite the white paper – and then hold that side up close to the flame of a candle in the dark. And then on my screen is an image of the candle-light – but hundreds of times larger than the needle-hole through which it passed. The light has not been in the least concerned by the smallness or shape of the opening – the whole thing has easily got through. But it is upside-down on the screen. The reason for this also explains the reason for its being there at all. We must never think of light as a stationary thing, but as straightlined rays advancing from every bit of the object at which they are being scattered. Thus the image is bound to be upside-down, for (see diagram) the arrow passing its rays through the hole must send them as shown. Instead of thousands of rays I have shown exactly two, the last ray from the extreme point of the arrow, and the last ray from the extreme end of the tail. If a screen obstructs the rays before they get diffused, and the hole is smaller than the arrow, then the point of the arrow is bound to be at the bottom instead of the top.)

Light, then, being a succession of waves in swift movement, enters the eye at the rate of so many million vibrations a second. Let us go back to the first primitive movement. It was made in response to chemical action upon the light-sensitive single-celled body. When the creature evolved more cells some responded to light while others were alone capable of movement. A communication cord linked them. This cord, or nerve, conducted an electric current as a message from the light-sensitive part demanding muscular expansion and contraction in the cells capable of movement. So what we have got here is – first, the

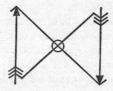

vibrations beating upon the eye, and second the nerve-conductor which communicates the message to the third thing, a muscle ready for action. As the animals developed in the course of evolution a networks of conductors came into existence, and the centre of this system, the brain, sorts out the messages and governs instantaneous action.

We use a term for this, a well-worn term now – we call it reflex action. It implies something purely mechanical. *It promoted a philosophy of mechanism.* It seemed possible to be able to eliminate the necessity for a spiritual principle. The fact that I see a red tulip in the garden can be given a purely physical explanation. I am not really seeing the tulip over there: I am not really looking at it – there is simply an image of a tulip in my head. How did it get there? It was excited into existence by vibrations or agitations working upon the material substance of my brain; these motions were set going by previous vibrations on the optic nerve; these by previous vibrations in the eye; and these by previous vibrations in space.

That is the upshot of the policy. You may feel inclined to say that this is absurd and that I am tilting at windmills. True, it is against the experience of mankind. But that does not prevent learned men from making the deduction and proposing a philosophy of mechanism. It is one of the many clever ways of being stupid. And just as thousands of modest men are easily floored by humbug and cant, not being able to detect these as such, so their sense of reality is easily clouded by determinists. A great physiologist once said to Ruskin that sight was 'altogether mechanical'. Ruskin comments that the words meant, if they meant anything, 'that all his physiology had never taught him the difference between eyes and telescopes. Sight is an absolutely spiritual phenomenon; accurately and only, to be so defined: and the Let there be light, is as much, when you understand it, the ordering of intelligence as the ordering of vision.'[16] Any ordinary person with an unclotted and inquiring mind (such as my own if I may say so) who goes to Optics to find out what *luminosity* is, why it is that we see *light*, gets no answer; nor does he receive any answer as to how vibrations can be ordered into images unless there is an agent inside our heads capable of doing it – a Mind, a Spiritual Apparatus at work over and above the machinery. 'It is the appointment of change,' Ruskin says again, 'of what had been else only a mechanical effluence from things unseen to things unseeing, from stars which did not shine to earth that could not receive; the change I say of that blind vibration into the glory of the sun and moon for human eyes.'

This may seem fairly obvious. Yet the melancholy fact remains that many of us when we are shown a mechanism conclude that it is only a mechanism. It is as if on my showing some one a watch he were to say that it had no spirit. But it has a spirit. It is bound to have a spirit in the form of a general maker and winder-up. The tendency to emphasize the mechanical at the expense of the spiritual seems to be a perennial danger and existed long before the modern age. 'If now someone should ask you,' says Socrates to Theaetetus, 'by what does a man see white and black objects and by what does he hear shrill and low

16. *The Eagle's Nest.*

sounds, I suppose you would say, by eyes and ears'. Theaetetus assents and Socrates continues, 'A careless ease in the use of names and expressions without pedantic linguistic analysis is for the most part not ignoble. The opposite is rather the mark of a mind enslaved. But sometimes it is necessary to be more exact. For example, it is necessary now to consider in what respect the answer you have given is incorrect. Reflect: which answer is more correct, eyes are that *by* which we see, or that by means of which *we* see, and ears that *by* which we hear or that by means of which *we* hear?' Theaetetus answers, 'It seems to me, Socrates, truer to say, *by means of which* than *by which* we perceive in each case.' 'Yes, indeed,' says Socrates, 'for it would surely be a terrible thing if so many powers of perception were seated in us like warriors in wooden horses, and if all these senses did not draw together into some one form, call it soul or what you will, by which, using senses as instruments, we perceive whatever we do perceive.'

It was necessary to say this in the days of Plato. It is necessary to say it today; for commenting upon that very passage, Dr H. G. Wood in the 1957 Eddington Memorial Lecture, adds:

I cannot see that this analysis is outmoded, or if it be outmoded I think it is still sound. It is truer to say that we see by means of our eyes than that our eyes see. And though the relation of ourselves to our brains is more intimate than the relation of ourselves to particular powers of perception such as eyes and ears, yet in this case, if the question is put, Do our brains think, or do we think by means of our brains? I should plead for the second alternative.

It was necessary for Coleridge to say the same thing before him. It was necessary for Goethe to say it at the time when it seemed that the new science was making it plausible to refer everything to mechanical and chemical reactions, and the perception of colour to be merely an automatic response of matter whether external to man or within the eye and brain. He may have spoilt some of the force of his affirmation by dogmatic assertions and physical caveats; he may have over-stepped the mark in speaking of Newton's 'empirico-mechanico torture chamber' and in declaring that he must rescue mankind from 'the pathology of experimental physics'; yet he was surely right in accusing the physicists of the day of subjectivity, of 'an excessive emancipation of the subject from a totality of which it is a mere part';[17] he was surely right in conceiving Mind as something in us not wholly *ours*, independent of whether we want it or not, a part of Nature playing its role within us as upon a stage, and subjective only in as much as it makes its appearance within us and thinks through our help; and he was surely right in repeating over and over again the dictum of Plotinus: 'If the eye were not sun-like how could we ever see light? And if God's own power did not dwell within us how could we delight in things divine?'

Goethe loved to personify Nature – quite in the old pagan manner. His response was such that he could not avoid it. Even when a boy, after listening to too much theological discussion amongst his family he would go to his bedroom, make an altar from a music-stand, deck it with minerals and flowers, and crown it with a flame lit by a burning-glass from the rays of the risen sun. His apostrophes to Nature are well-known. It is not strange that such a man would react as violently as Blake to the abstractions of Descartes and Locke. We must remember that these poets were eighteenth-century men reacting to the new thing. The first reactors are always the most violent. Think of the violence of the first reactors to industrialism!

Yet there seems to be confusion as to the final upshot of what was being put forward. In an often quoted passage, Whitehead outlines the theory of primary and secondary qualities as proposed by Locke and colour or scent or sound are secondary qualities not actually belonging to the given mass. Then he goes on:

Mental apprehension is aroused by occurrences in certain parts of the correlated body, the occurrences in the brain for instance. But the mind in apprehending also experiences sensations which, prop-

17. Erich Heller's 'Goethe and the Idea of Scientific Truth' in *The Disinherited Mind*.

erly speaking, are qualities of the mind alone. These sensations are projected by the mind so as to clothe appropriate bodies in external nature. Thus the bodies are perceived as with qualities which in reality do not belong to them, qualities which in fact are purely the offspring of the mind. Thus nature gets credit which should in truth be reserved for ourselves; the rose for its scent: the nightingale for its song: and the sun for its radiance. The poets are entirely mistaken. They should address their lyrics to themselves, and should turn them into odes of self-congratulation on the excellency of the human mind. Nature is a dull affair, soundless, scentless, colourless: merely the hurrying of material, endlessly, meaninglessly.

However this may be disguised, says Whitehead, this is the practical issue of the scientific philosophy which closed the seventeenth century and was carried on into modern times. 'And yet,' he adds, 'it is quite unbelievable. This conception of the universe is surely framed in terms of high abstractions, and the paradox only arises because we have mistaken our abstraction for concrete realities.'[18]

Apart from the fact that probably not one single person has ever taken the abstraction for the concrete reality, there seems some lack of fusion here – at least I find it confusing. For that philosophy, if taken on its own ground, should not upset anyone. Not that Mind is by no means excluded. On the contrary it is represented as doing a big job. If we strip the passage of its rhetoric we find that Whitehead is simply saying that the mind is largely responsible for harmonizing our stream of impressions into a glorious picture. Now if this is really so and that we thus make sense out of otherwise senseless vibrations, why should the poets not do us some odes upon the excellence of the human mind? We still have the thing in itself, the finished article.

Goethe said that it was no use trying to wrest the truth of Nature by using the screws and levers of science – we can only get at the facts that way. To get at the truth, the significance, the quality, we need another instrument. The spanner we need for that is imagination. Science deals with how things are as-

18. *Science and the Modern World.*

sembled, imagination with the value and significance of the assembly. The same man ought to be able to use both tools. When we use the latter we do not abstract all the vibrations, we contemplate the thing itself: and then – what a difference! Colour, for instance, seems to have nothing to do with theory or controversy, and shines before us as the most powerful and the most redemptive of all the servants of beauty.

6. Enlightened Men

We come to a final consideration. Literature and Scripture is riddled with the symbolism of Light. Hymns, sermons, poems, rituals, exhortations, prayers are laden with metaphors of light. In fact there is too much of it: we weary of so much illuminated metaphor. Sometimes it is used as more than metaphor. When it is said that God is love it is a statement rather than a metaphor. So also with God is light. The terms seem almost interchangeable. 'Since God is light,' said Milton, 'And never but in unapproached light Dwelt from Eternity' – it was 'Bright effluence of bright essence increate.' We turn to the religion of the Zoroastians who were said to take stars for money rejoicing at seeing something they could not put into their pockets: for them the vital principle of their religion lay in the recognition of one supreme power, the God of Light, in every sense of the word. We turn to the Koran. 'God is the *light* of the Heavens and Earth,' runs Sura XXIV. 'Light is like a niche in which is a lamp – the lamp encased in glass – the glass, as it were, a glistening star. From a blessed tree it is lighted, the olive neither of the East nor of the West, whose oil would well nigh shine out, even though fire touched it not! It is light upon light.' We turn to the Bible. 'God is light and in Him is no darkness at all'; 'I am the light of the world'; and of course the many exhortations – 'Walk in the light', 'Let your light so shine before men', 'Believe in the light', and 'In thy light we shall see light'.

It scarcely matters whether all this is metaphor or not, since a metaphor is such a profound thing. Light is the term used to denote spirit. It is also the term used to embrace matter. 'We have come into the possession of a wonderful principle,' says Bragg, 'which unites all forms of radiation and all forms of matter. We may rightly speak of light as constituting the Universe.' Certainly religion and science come very close together here. The outer light and the inner light may be different aspects of the same thing. Since this would be merely an intellectual synthesis I do not know that I would bother to mention it were it not for the existence of a more compelling sign concerning the spiritual nature of light. I refer to the illumination – in two senses – which so often accompanies the mystic experience.

The mystic experience. This is surely the only ultimate. A mystic, however much and however often the word may be abused, should be defined as the man who is no longer mystified – by religions, by theologies, by doctrines, by formulations. He is at ease with religions, having attained Religion: he is no longer diseased by overdoses of doctrine. He occupies this favourable position by virtue of his experience which brings with it such joy and such certainty that he no longer needs dogmas about salvation: this experience to him is salvation. He no longer needs to solve the lesser problems: they have been dissolved. It may not last for more than a few minutes but it changes the man's outlook for ever – *he has seen the light*. These men are our true leaders; their goal of higher consciousness our chief hope. Their message comes down the years always with the same power to convince, always undated and undating. Their words bring continual support to the whole and the happy, and they bring hope and healing to the unhappy, to the broken in heart and even in purse, to the doubtful and to the dying.

This crucial experience is often accompanied by intense inner illumination. Yet the accent is not on the inner but on the illumination: it just is *light*. It matters little where we turn for examples since whatever source we take them from they are equally convincing. 'Light untellable,' says Whitman, 'lighting the very light' – he who, unlike most mystics, dwelt for long periods on this plane though in a quiet way. George Fox, terribly compelled, indeed propelled, by his inner voice, tells how he found himself immersed in an ocean of light. We know how Wordsworth expressed it, as the poet's additional gleam, the consecration, 'the light that never was on sea or land.' We recall the simple words of Havelock Ellis – 'I trod on air, I walked in light.' We think of the pathetic and terrible cry of A. J. Symonds – 'It is too horrible, it is too horrible, it is too horrible!' For while he was under chloroform he had an ecstatic vision of God. 'I felt him streaming in like light upon me . . . I cannot describe the ecstasy I felt.' Then he awoke from the influence of the anaesthetics and the vision went. It was intolerable. He could not bear it. 'Why did you not kill me? Why would you not let me die?' he shrieked at the frightened surgeons. He felt he had been tricked. Since the psychic experience had come to him through the physical loosening caused by the drug, he thought, as a man taking mescalin might think, that the experience itself was therefore suspect.[19] In *My Quest for God* J. Trevor tells how on one brilliant Sunday morning he felt he could not accompany his wife and children to chapel but must remain on the hills – he felt that it would be spiritual suicide to go down. So, reluctantly and sadly, he parted from his wife and boys and went further up into the hills with his stick and his dog. 'In the loveliness of the morning, and the beauty of the hills and valleys,' he says,

I soon lost my sense of sadness and regret. For nearly an hour I walked along the road to the 'Cat and Fiddle' and then returned. On the way back, suddenly without warning, I felt that I was in Heaven – an inward state of peace and joy and assurance indescribably intense, accompanied with a sense of being bathed in a warm glow of light, as though the external condition had brought about the internal effect – a feeling of having passed beyond the body, though the scene around me stood out more clearly and as if nearer to me than before, by reason of the illumination in the midst of which I seemed to be placed.

In his book called *Cosmic Consciousness* which has been an inspiration to many, R. M. Bucke describes how he returned

19. *The Varieties of Religious Experience.*

home one evening after talking with friends. He says he was in a quiet and passive mood.

All at once without warning of any kind I found myself wrapped in a flame-coloured cloud. For an instant I thought of fire, an immense conflagration somewhere close by in that great city; the next I knew that the fire was within myself. Directly afterwards there came upon me a sense of exultation, of immense joyousness accompanied or immediately followed by an intellectual illumination impossible to describe. Among other things I did not merely come to believe, but I saw that the universe is not composed of dead matter, but is, on the contrary, a living Presence.

The utterances of these enlightened men brings me to the end of this book on light, which, starting with physics, ends quite naturally with metaphysics. In between I have described certain phenomena, hoping that though little can be explained about any thing, we may come to see that it is – and we can scarcely go deeper than the Biblical words, *I am that I am*. Perhaps we solve the riddle of the world by being able to see the world. Why lose our one chance to see it? What shall it profit a man if he lose the whole world without any guarantee that he has gained his soul? If we use *all* our faculties, little will seem common-place, much may be transfigured. I have sought to do no more than take a mirror, wipe it, and place it in your hands.

BOOK II
THE MOVING WATERS

My aim at sequence here has been much harder to achieve. Not only have I omitted the whole of the first Chapter on The Firmament, but have had to omit some of the 'The Vestures of Water' – at the beginning, 'Hail', and at the end, 'Dew'. And in my determination – I hope not misguided – to fit in passages which I am unwilling to leave out, I have been compelled to put dots to indicate previous pages which have built up my theme to that point.

Still, if this results in lighter reading, perhaps no one will mind except myself!

– JOHN STEWART COLLIS

CHAPTER I
SECTIONS FROM *VESTURES OF WATER*

1. Snow

... Hail, then, is frozen rain. This is not the case with snow; it does not take its origin from any form of rain. The birth of snow is very beautiful: we shall not encounter a more delicate miracle, nor attend at a deeper mystery. Hast thou entered into the treasures of the snow? asked God of Job, and Job was silent. Century follows century and still most of us are silent about this, as about all such things. No, we say, we have not entered in; we have not the interest, nor the strength, nor the devotion, nor the love, to get to *see* the images and read their scripture. Must our answer for ever be one of shame, or shall we try, shall we try now to enter into the treasures of the snow?

When we take in our hands a given substance – say, a diamond – we regard it as a unit complete in itself. We do not see, for we cannot see, that it is really a compact grouping of millions of invisible units which have come together and formed it. Yet such is the fact. We call these invisible units molecules. Some of them show a far greater tendency to come together and form a solid than others; and all of them will resume their individual independence and abandon their unity under the influence of intense heat. At one time the earth was so hot that it contained no solidity whatever and there was nothing to be seen save gases and fluids. It cooled. And as it cooled the various sorts of molecules adhered into different mineral formations.

Why did they do that? Because of the nature of their remarkable properties. These molecules, these particles of matter, though so small as to be invisible individually, possess attractive and repellent poles. An attractive pole is a magnetic point. A magnetic point is an area of matter – however minute – which

pulls other allied bodies towards itself. Thus in approaching
snow we come to magnetism and draw close to one of the final
mysteries. I would not, even if I could, go further now into this
and say what this attractive power *is*, passing from cause to
cause and seeking ever further shores to find the father; it is
enough for me at the moment to have the privilege to gaze with
awe through this one doorway at the dance of life. There are
these poles of attraction and repulsion. Under the influence of
heat – a deeper mystery – they dart about so rapidly that in spite
of their mutual attraction they do not cohere. But when the
temperature cools their motion becomes slower and they begin
to come together, and according to the measure of their cooling
is the measure of their assembly. And here is another marvellous thing. They come together; they form solids: yes – but
in no haphazard manner. They build themselves into definite
and lovely forms: we attend at the birth of symmetry and geometry. We call them crystals. Apply intense heat to iron or
copper, to gold or silver or lead: let these solids loose and flow
in fluid shape: then lessen the heat and you will see the crystals
grow. Out of the fluid flint comes the rock crystal; out of dissolved chalk comes Iceland spar; out of particles of silica the
amethyst; out of carbon the diamond. We look round at our
world today: we stoop down and pick up the precious stones,
the ruby and the sapphire, the beryl, the topaz, and the emerald:
we see them glitter and we see them shine – but yesterday they
were cast into the oven of the earth and flowed in liquid
form.

The air is full of the molecules of water vapour. Their constitution is such that in the normal temperature of the atmosphere their poles are pushed so far asunder as to be practically
out of each other's range. They do not unite. They freely flow in
their swarming millions through the air. But when the temperature begins to freeze, then their motion is slowed down and
their mutually attractive power, their magnetism, draws them
together, they become visible, and we see crystals made of ice.
And again, the points of attraction and repulsion belonging to
each particle exert such an exact degree of pull and pressure

that the most delicate shapes of marvellous symmetry are built up. These are the flowers of the sky. We call them snow.

The crystals come about as I have described, but there is a further point to notice about their architecture. They need a substratum upon which to grow, a supporting substance as a foundation. They get this from the dust in the air. We might not think so, but there is a vast amount of it, for not only is there earth-dust consisting of pollen, bacteria, soot-spores, volcanic ash, and so on, but there is sea-dust, and star-dust from outer space which is said to fall to earth at the rate of 2,000 tons a day. We do not notice it normally, but we do see it most distinctly when it is caught in a sunbeam. In order to satisfy ourselves that the dust in the air is not simply an urban effect, we need only go out into the country on a half-cloudy day when sunbeams slant from the sky. Then they reflect the dust so clearly that we see great shafts, brighter than the light around, pointing down from a glowing gap in the clouds like searchlights reversed, often passing from field to field as if looking for something.

It is upon this foundation that tiny discs or needles of ice are built – and upon them that the finished article, the crystal, is displayed. We cannot see the work: can we imagine it? 'Let us imagine', said John Tyndall, 'the eye gifted with a microscopic power sufficient to enable it to see the molecules which composed these starry crystals; to observe the solid nucleus formed and floating in the air; to see it drawing towards it its allied atoms, and these arranging themselves as if they moved to music, and ended by making that music concrete.'[1] This concrete production is truly marvellous in the artistry of its geometrical design. The variety of appearance is inexhaustible, but very often (though it would be rash to say always) the foundation of an hexagonal shape is adhered to, so that each is a little star with six rays crossing at an angle of sixty degrees. Then if the crystal looks like a composition of ferns it will have six out-pointing leaves; if like a windmill, it will have six sails; if like a sundial, it will have six corners; if like an assembly of

1. *Forms of Water.*

swords, it will have six blades; if like a star-fish, it will have six ribs; if a fir tree, it will have six stems with plumes likewise set in perfect symmetrical precision.

I write these facts down, for it is proper to do so, but it can scarcely be of much help to the reader. Here is an art which cannot be dealt with through the medium of words, it would be unwise to make the attempt. The right department is not literature, but photomicrography. It is only within the last seventy years that this scientific art has come into play and revealed the snow crystal to us in all its glory. We think at once of Wilson Bentley of the United States. He was a saint of snow. He devoted his lifetime to its study. From the day when his mother showed him a snow crystal through a microscope he dedicated his life to that service. At first he sought to copy with pen and pencil what the microscope revealed, but he had not time to trace their delicate beauty before they had perished. But in 1885 when he was twenty he acquired a photomicrographic camera, and for forty-six years from then until his death he photographed snow in his lonely mountain retreat in New England, taking over 6,000 negatives of crystals. In 1931 his great work, *Snow Crystals,* was published, his task was finished, and so he died. Truly he has enabled mankind to enter into the treasures of the snow.

Bentley was unable to discover, out of his 6,000 pictures of snow-crystals, two which were identical. It has been calculated – by what means I know not – that in an average-sized snow-storm something like 1,000 billion crystals descend. In the boundless prodigality of her beauty Nature rains down these starry gems, these fleeting flowers of aerial ice – each one different. I mention this because it is a lovely thought. But it is not a surprising thought. It would be surprising if any of them were quite identical. People sometimes say: 'How extraordinary that we never see two clouds exactly alike' – or two trees, or two faces, or anything else. Yet it is not extraordinary. Put yourself at the creating end. You are given a certain fluid material with which to make a cloud: how frightfully difficult so to fix the vapour that it will ever work out into exactly the same shape. How much more so with the complicated materials that go to make a face. The same with snow crystals: it is a wonder of wonders that the dance of the magnets produces these geometrical designs – it would be too much to ask that the exact same shape in every particular could ever be duplicated.

I have just said that in a snow-storm so many billions of crystals descend. But only in very rarefied places and altitudes do they come down singly. Many of us are not particularly conscious of the existence of the tiny, delicate ice filigrees because in warmer places they come down in bunches. We call these aggregates snowflakes. Sometimes they look like feathers, sometimes like the ghosts of leaves. And when they alight, falling one upon another, layer on layer, still less do we see them individually – they have become banks and fields of one white substance. Yet is it white? Not in itself. It is frozen water, colourless ice, and the whiteness is produced by the reflection and refraction of light from the myriad minute surfaces of the crystals. However, whiteness is what we see, and we are appalled or purified according to our mood and to the scene. Those smooth curves, those far high lands beyond our reach, those intoxicating slopes that have given wings to our feet and wine to our heads, those heartless voids and weary wastes of dumb blankness full of menacing meaning – how many are the images of snow that hang in our memories. All given their character, sublime or pitiless, by the invisible play of molecules.

The crystals fall upon the earth, delicately, weightlessly. In the end there is weight, a great deal, under their massive multiplicity. House-roofs can be caved in and ordinary trees become weeping willows; but the building up of their load is the very symbol of lightness of touch. It has snowed all night, and in the morning we enter the wood. The evergreens are holding up thick fists of snow like white boxing-gloves, and the fingers of other bushes are wearing batting-gloves. It is curious to see long twigs, thin as wire, serving as the base for quite high walls of snow – a thing to be declared impossible were it not before our eyes. The laying of the bricks has been so gentle and each in itself is so light that a perfect balance is achieved, and the pure white snow on the clear black boughs is both heavy and light.

It alights so delicately that a great amount of air is present in newly fallen snow. Extreme cases have been given of drifts composed of one part ice to eighty-nine parts air. For this reason men who have been buried deep in snow for as long as two days have not been suffocated. And for the same reason it is a positive aid to agriculture. It serves as an excellent insulator. Charles F. Brooks in his *Why the Weather* makes the interesting observation that an experiment showed in a given area that the temperature of the air three feet above the snow was minus 19° Fahrenheit, and at the surface was 27° Fahrenheit, while the temperature of the snow seven inches under the surface was 24° *above* zero. Thus seven inches of snow between the surface and the interior of the snow covering raised the temperature 51°. And on another occasion he was able to record a difference of more than 60° between the top and bottom of the snow cover. Hence the temperature of the soil is kept uniform. The white blanket keeps heat in the lower ground levels and preserves seeds from freezing to death in excessive cold. When the snow melts and the blanket is lifted there will be no trace of frost. 'This absence of frost', remarks Brooks, 'has an important influence in preventing spring floods. The water of the snow melting in the warm sunshine of approaching spring is absorbed freely into the soil. As a result pastures and lawns may emerge from the snow cover verdant with young grass.'

Yet that is only one side of it, of course, and the snow crystals can scourge us with blizzard and avalanche. We do not think of pretty crystals in a blizzard; we face a gale of wind, zero cold, hurling at us powdery snow. A blizzard is not something descending from above; there need be no snowfall, yet to the height of seven hundred feet the air may be filled with swirling masses of snow, whipped up by a gale. You can fly above it in an aeroplane in clear blue sky. Avalanches are more dramatic. So uncertain is the clasp of the huge, snowy hands that cling to the mountains that the least motion may unsettle them and send thousands of tons of snow crashing to the bottom of a valley and even halfway up an opposite slope. The vibration of a train, the fall of a branch, the passage of a fox, the ringing of a bell, the crack of a rifle, the voice of a man, even the passing of a cloud, have all at times been sufficient to set the vessel loose and make it take the plunge. I once witnessed one from the Mer de Glace. It was as if a white cumulus cloud had somehow lost its balance and fallen from the sky into a ravine, down which it rolled from rock to rock in smoking ruin. And it made a noise like thunder – thunder too had fallen from its high estate and issued from that gloomy gorge.

Avalanches seldom fall upon the habitations of men. The Sicilian peasant, for ever faithful and for ever spurned, still clings to Etna's side. His villages and vines, his orange groves and wayside chapels, may disappear beneath the lava, they have indeed often done so, as if sunk in the sea – but he returns. Few villages are built within the reach of avalanches. But sometimes they do descend upon the ways of men with apocalyptic disaster. Then there is no escape.[2] Lava creeps at only so many inches an hour as it arrives below, but the avalanche sweeps all before it, as, for example, when in 1910 one struck Wellington in the Cascade Range, killing ninety people, while three large locomotives, some carriages, a water-tank, and the station building were swept over a ledge into a canyon 150 feet below. And with the fury of the fall there is added another terror – the blast. Extraordinary velocities of air are set up by the headlong advance of the snow-mass. Air becomes imprisoned and compressed within the middle of an avalanche, and on obtaining an outlet bursts into the atmosphere with explosive violence, creating a blast very like the familiar experience of bomb-blasts. Thus their effect is often spectacular, as in the case of an avalanche in Alaska, the blast of which hurled cabins 100 feet up a mountainside, and as on a famous occasion in Switzerland when an iron bridge weighing many tons was thrown upwards 150 feet; while there is the classic story of the Alpine stage-

2. The happenings in Switzerland in the winter of 1953 add a tragic postscript to this.

coach which was thrown clean across a river, together with coachman, horses, and travellers.

The most pleasing example of human crisis brought about by the invisible play of the magnets at their musical dance in the aerial halls is the snow-storm that fell upon New York on 12 March 1888. After one night's fall the city was half-buried. The lights failed. The traffic stopped. Even the elevated trains came to a halt, and the occupants descended by ladders. 'The city', wrote Free and Hoke,

was flatly conquered by the weather. For two days it was out of touch with the world. No street cars ran for four days, nor could food be delivered. Rail and wire communication was cut off. Normal train service was not restored for ten days. Millions of dollars disappeared in lost business and ruined equipment. Hundreds of thousands more it cost to dig out the streets and railroads. And it took the snow only six hours to do all this damage.[3]

It is good to think of these things. It is encouraging to attend at the inexhaustible power of Nature to create, to destroy, and to paralyse. It is comforting to feel that we are often helpless when she strikes, for hers also is the hand that holds us. I love to think of one particular office which she can perform with snow. See those falling flakes! Are they not also the messengers of silence and of peace? Of *silence*. The feeble filigrees, the fairy ferns descend: down they come, the bonny gems, as if emptied from some treasure-house above. Each is so little. They become so much. Here is our city. Here our roads. Our civilization is in full swing. The great principle of moving from one place to another place at all costs and as quickly as possible is working well; the absolute necessity of combustion as the main rule and law is heard throughout the land and shrieked throughout the street. And the white flowers fall – and still they fall. They stick and stay where they have lain. They lie one upon another; now an inch of them along the road, the street; then a foot; many feet at last. A hush; a pause; as soundless fall the feathers. See, the roads are empty and combustion wanes! Civilization has

3. Free and Hoke, *Weather*.

been stilled. Silence has fallen upon the roaring wrack around. These the frailest of all earthly envelopes, and perhaps the most beautiful, have yet the power to bestow peace upon the world.

2. Ice

If the crystallized formation of snow is not immediately obvious to us, still less is the similar architecture of ice. Just as the snow-flowers are constructed by the molecules of vapour in the calm, cold air, so also a six-rayed star is typical of the construction of all lake ice; but ice and water are so optically alike that unless the light fall properly upon these figures we cannot see them. They are there in myriads, sometimes so small that we need a magnifying glass to discern them, all wonderfully interlaced. The ice-covering of a lake is a carpet of rare design, far beyond the power of man to produce.

But how comes it that there should be that carpet or floor frozen on *top* of the water? We would expect it to sink to the bottom.

I take a stone in my hand and throw it into the water – and down it goes to the bottom. I take another stone in my hand, of exactly the same weight, but made of ice, and throw it into the water – and it floats.

I have thrown a lump of water into water, and it floats. A water-stone has become a boat. This is certainly curious. It doesn't float because it is porous: it is not like a piece of wood, with air in it. It feels like a slippery stone, and is the same weight as the stone which I have just thrown in – but it does not sink. Yet, according to the rules, it should sink.

The application of cold, we have been led to expect, does not expand things; it contracts them, it makes them heavy. Warm the air and it will rise – like your heart: cool it and it will descend. Hot water rises and cold water descends. Why then, if water becomes so cold that it actually congeals into a solid, does it not at once go to the bottom? We may note, in passing, how very convenient this is. Were it not so the lake would gradually fill up with ice, become a solid block like a glacier, with not enough heat to melt it at any time, while the fish would perish, and in fact the whole economy of Nature as we know it, with its free flowing of water throughout the earth, would be, not just upset, but overthrown. But instead of sinking it stays on top, creating a protective roof, as if with a special eye to the preservation of fish and the world's comfort generally.

It does this for the same reason that a lump of iron will not sink in a pool of liquid iron. If under the pressure of intense heat iron is melted into fluid – the embrace of the molecules having thus been loosened – and then a piece of iron is thrown into the molten pool, it will float.

Thus solid water is not peculiar in this respect. The explanation is subject to a clear statement, I think, though perhaps my way of putting it may be too simplified. Again then, as we know water is composed of myriads of molecules. I take a cupful of water and put it in a freezing atmosphere. I find that when the ice has formed it requires more room than it occupied in the cup at first. It has expanded. The given number of molecules now occupy more space than they did in the liquid form. And because they occupy more space it follows that this slab of ice contains fewer molecules in it (fewer bricks to its building) than that amount of fluid space: so it is lighter than that amount of fluid space and floats upon it. We must remember that by lighter we simply mean that there is less substance there to be pulled down by gravity; and since each specific substance has its specific gravity the flint-stone will be pulled down while the ice-stone is not pulled down. And if the specific gravity of any stone I throw into a pool of molten iron is greater than the specific gravity of iron, then it will sink while the iron-stone floats.

This answers the first part of the question – why the ice floats. But now we must see why the molecules expand instead of contracting under the pressure of cold. They do contract, they do draw closer to one another at first. But we have to reckon with the activity of the magnets in the molecules. When the molecules begin to come together under the influence of cold, then the polar forces come into play. 'The mutually attracting points close up, the mutually repellent points retreat', says John Tyndall, 'and it is easy to see that this action may produce an arrangement of the magnets which *requires more room*. Suppose them surrounded by a box which exactly encloses them at the moment the polar force come into play. It is easy to see that in arranging themselves subsequently the repelled corners and the ends of the magnets may be caused to press against the sides of the box, and even to burst it if the forces be sufficiently strong.' In short, the water, when becoming ice, does at first contract in volume under the influence of cold, but later expands in volume because, under the influence of the attraction and repulsion of magnetic forces, the molecules are compelled to turn and re-arrange themselves, demanding, as they do so, more space. Thus the expansion.

The magnets demand more room. They demand it, and they get it. In order to prevent pipes from bursting under their pressure, at 20° of frost they must be strong enough to resist 138 tons to the square inch. We call the mending of broken pipes plumbing – how such work must fire the imagination! We have the visible, hard, concrete pipes: inside them the soft, fluid substance: composing it the invisible particles – and they have the power to burst the vessel. Well, we live in an atomic age. We have realized by now that the principalities and powers do not advertise themselves in outward splendour, they are not found in gorgeous palaces – neither spiritually, politically, nor physically.

3. Glaciers

Though lakes and seas do not freeze solid, on account of this same law, there are plenty of frozen blocks covering wide areas. In fact 6,000,000 square miles of the earth's surface is permanently covered with ice. The continental ice-sheets of Greenland and Antarctica are the remnants of the great ice-masses which at the beginning of the Quaternary Era covered so much of the earth that we speak of an Ice Age. In those days much of the oceans lay blocked on the land, and it is calculated that even today, if all the ice-sheets went back to the sea, it would rise 150 feet and submerge a large portion of the land.

Five-sixths of this 6,000,000 square miles are occupied by what are called ice-caps covering plains and mountain plateaux in the polar regions. They can be, as in Greenland, 6,000 feet deep. That 6,000 feet depth of ice is not a question of a lake filled in – as it could not be; it is snow which has turned to ice: it is a glacier. The remaining million square miles are occupied by sloping glaciers. I propose to say a word about these because they are interesting in themselves and important in the economy of Nature.

Although I have called the atmosphere a kind of greenhouse, its greenhouse quality can really be applied only in terms of feet above the surface of the earth. It becomes freezing cold so soon that there is no rain upon the higher reaches of the high mountains – only snow. On such mountains there is what we call a snow-line, which is the line above which the snow never melts and below which it is completely cleared away every year. Now, if the snow above the snow-line were augmented every year without any melting it is evident that the summits of the mountains would grow; they would go on going up, piled higher every year, not with stone but with solid ocean. Two things prevent this from happening: first, continual avalanches occur as the building becomes top-heavy (as it is bound to do); and, second, numerous long, compacted blocks of snow, called névé, which by pressure and freezing become ice, descend into the valleys. These are the mountain glaciers of the world.

At its bottom the glacier melts and serves as the source of a river, such as the Rhône or the Arveiron. If it is in equilibrium the amount of body accumulated at the upper end will equal the amount melted below. A glacier is therefore a fine reservoir that stabilizes the flow of rivers which would otherwise be subject to violent fluctuations.

If the glacier melts at the bottom, and is fed from the top, then it must move. That great winding mass of solid ice must be in *motion*. This seems contrary to our senses, if not our sense.

We climb on to one: it is not moving: it cannot move. So we think, and it is only within the last 100 years that scientists ceased to deny the idea and to declare that it was manifestly absurd. Today it is a recognized fact that glaciers do move. Their various rates of movement have been measured, and the fact established that they move faster in the middle than at the sides. This second fact was even more perplexing than the first. It was hard enough to credit that this high mountain road, as firm as any other road, but with deep clefts called crevasses often running across it, was sliding down the hill; but it was still less easy to believe that it did not move all in one piece, that the middle of this rigid block moved faster than its sides, and that it did this without the crevasses closing up, which surely should then be found closing and opening like a concertina. How could a substance as brittle as glass flow like treacle? In due course, however, this also was accepted, and it was agreed that ice is not a solid but a fluid, and flows down the hill like any other fluid.

I set down these facts in comfort and with sobriety. But the establishing of them meant a hundred years of devoted experiment and bitter controversy among the geologists who opened up this particular branch of glaciology. Scene I: Does it move? Enter Professor Hugi. He builds a hut upon a glacier above the source of the Arveiron, to make exact observations. His hut moves. From 1827 to 1830 it moves 330 feet downwards. By 1836 it has gone down 2,354 feet; by 1841 it is 4,712 feet below its first position.

Scene II: Enter Louis Agassiz. He takes shelter on the same glacier under a great slab of rock with some companions, and they add sides to it and call it Hôtel de Neuchatelois. The hotel goes down. In two years it has moved 486 feet. Thus the main fact of movement is established – though I may add here that nothing which these people did offered so striking a proof as that given later by Captain Arkwright, who in 1866 got stuck in a glacier and went down from top to bottom, a distance of 9,000 feet, which took him thirty-one years.[4]

Scene III: Now for greater accuracy in measurement. A very

4. See Aubrey Le Bond, *True Tales of Mountain Adventure*.

obvious sort of instrument was invented which they called the theodolite. Now it was no longer necessary to wait a year before recording results; motion could be calculated from day to day – nay, *watched* from hour to hour; each glacier having a different rate of advance, according to its decline, from so many inches to so many feet a day.

Scene IV: Enter J. D. Forbes. Glacier motion has been accepted, but ice is still thought of as an entirely solid substance, capable of being broken or crushed or pushed or pulled, and of falling or sliding, but always rigid and brittle like so much stone or glass. Forbes proceeds to put an end to all that. Observing how the dirt-coloured bands of debris spread out over the glacier in parabola shape, he recognized the essential fact that the flow was greatest at the middle. This lead to the classic book on the subject, his *Travels Through the Alps*, in which he proved that glaciers are not solid bodies, but semi-liquid ones, and that they run down their beds like treacle.

Scene V: Resultant mortification of other glaciologists, who, though longer on the job, had failed to see the obvious. The baser elements of human nature now exemplified by the weaker brethren. Great annoyance of Agassiz, who wanted the credit. Vituperation against Forbes. The viscous quality of ice declared 'not proven'. Ten years' wrangling and quibbling over the word 'viscous' and the word 'plastic' and the word 'elastic' and the word 'malleable' and the word 'ductile'; and disputes as to whether ice is more like ice-cream than treacle, or putty than glue, or lava than wax, or clay than butter, or jam than jelly. And so on, through Charpentier's theory and De Saussure's theory and Rendu's theory and Thomson's theory and Tyndall's theory, until we get Faraday's theory of regelation, which made everyone pleased by proving that blocks of ice which have been broken apart can be sealed together again. 'It is very interesting to know that if you put two pieces of ice together they will stick together,' said Ruskin.

Let good Professor Faraday have all the credit of showing us that; and the human race in general the discredit of not having known as much as that, about the substance they have skated on, dropped through, and ate any quantity of tons of – these two or three thousand years . . . I do not doubt that the wonderful phenomena of congelation, regelation, degelation, and gelation pure without preposition, take place whenever a schoolboy makes a snowball; and that miraculously rapid changes in the structure and temperature of the particle accompany the experiment of producing a star with it on an old gentleman's back. But the principle conditions of either operation are still entirely dynamic.[5]

In a work of this nature I am constantly on my guard against falling into the pit of past controversy and engaging the reader's attention on subsidiary problems of yesteryear. I have paused here – with almost telegraphic brevity – to mention this scientific controversy because it serves so well as a reminder that behind the simplest statements and 'ascertained facts' which we all take for granted on any subject there generally lies a great campaign with various schools of thought, carried out for years by remarkable men devoted to the truth – if not always to each other.

Let me round this story off. You may think that Ruskin adopted too lofty an attitude in that last quotation. But he knew his stuff. He was really a born geologist who had written a 'Mineralogical Dictionary' as far as C, and invented a shorthand symbolism for crystalline forms, before he was fourteen. He knew a great deal about the Alps; they were the greatest inspiration of his life (as they were also responsible for the most apocalyptic passages of Wordsworth and the four finest lines of Byron). He had studied the glacier question over a period of forty years, and in 1875 he summed up as follows: 'The first great fact to be recognized concerning them is that they are *Fluid* bodies. Sluggishly fluid, indeed, but definitely and completely so; and therefore they do not scramble down, nor tumble down, nor crawl down, nor slip down; but *flow down*. They do not move like leeches, nor like caterpillars, nor like stones, but like, what they are made of, water.'

When I turn to the latest monograph on the subject, 'Glaciers', in *Science News, No. 6*, by D. M. F. Perutz (who was a

5. *Deucalion*, Chap. III, 'Of Ice-Cream.'

member of the 1938 Jungfraujoch Research Party), I find nothing in it which contradicts anything Ruskin said on the subject; but we can go farther with Dr Perutz into the mechanics of how the ice does manage to get along by studying the theory of 'creep'. Creep does not mean that anything creeps. It means it shrinks. The laws of molecular motion that bring this about in glaciers explain the plasticity of ice, and 'show that no special mechanism – such as regelation – need be invoked to explain the existence of streaming motion in glaciers, since ice is only one among many other apparently brittle crystalline materials which can be deformed plastically by creep under suitable conditions' (one of the conditions being *pressure*). And, finally, it is understood now that crevasses, so far from suggesting that the glacier cannot be moving downhill, are on the contrary an indication that they are doing so, and that the more crevasses we see the greater is the speed of the descent, since they mean that the stretching of the ice has been under such a strain that it has cracked at various points, just as mud will crack and even jelly will crack; and since the said cracks or crevasses do not by any means go right through the block, the mass can move downward without causing them to close.

That is all I propose to say on this aspect of glaciers, though I fear that even so I have said nothing as to the manner in which the snow has turned into ice; for I cannot hope to conjure up before my own mind, let alone the reader's, the pressures, the squeezings, the squashing of the air-bubbles, the temperature changes, the tricklings of water within that never motionless gigantic mass where every crystal grain is involved from the time when it fluttered down to form the highway.

I use the term 'highway', for that is how I always see them, whether in pictures or in reality – highways not built with hands. I have never gladly contemplated one, and crevasses fill me with a horror I do not experience when looking over a cliff. Nothing exceeds the sadness of those streets. You can see them leading on and on ever farther into the high misty peaks, a vision of freezing terrible loneliness; a vision not just of Nature not only objective but subjective; a vision of our own private and appalling solitude: and that far, sad, lonely road is not leading to any home, nor any heaven – but, cold and comfortless, to journey's end.

Nevertheless, there is traffic along these roads. There is a gigantic work of transport in operation. Vast agricultural schemes are being carried out.

Second only in interest to the glaciers themselves are the things we see upon them. Over there, for instance, we see a flat slab of rock perched upon quite a high column of ice, like an imitation of a mushroom with a long stalk – it is called a glacier table. The stone is a parasol; it is up there, not by going up, but by not going down; the surface of the glacier has melted several feet all round, but not under the slab; so it has built its own pillar – from head downwards. And over there, on that promontory from which the glacier has receded, is a huge block of granite. It seems not to belong to its environment, but to have come from far. Wordsworth had some such block in mind in *Resolution and Independence* when the Traveller upon the lonely moor sees the Old Man beside the bare mountain pool:

> As a huge stone is sometimes seen to lie
> Couched on the bald top of an eminence;
> Wonder to all who do the same espy,
> By what means it could thither come, and whence;
> So that it seems a thing endued with sense:
> Like a sea-beast crawled forth, that on a shelf
> Of rock or sand reposeth, there to sun itself.

We find such rocks near glaciers, or where glaciers have been. They are called erratic blocks, for they have erred and strayed. What a river carries fast at the bottom of it a glacier carries slowly at the top of it, and often much farther, and often rocks which no river could carry a yard – such as that 24,000-cubic-foot stone at Mattmark Lake, deposited by the Schwartzberg Glacier. These erring stones have broken from summits far off, have fallen on to a glacier, and like a log upon a river have flowed away, in some cases journeying for two thousand or even four thousand years before coming to their monumental rest.

These glacier tables and erratic blocks provide the more outstanding examples of glacier transport. But they are less important than the positive rivers of debris, called moraines, which edge the glaciers, and when one glacier joins another become medial moraines, their line then being in the middle of the now combined ice-stream. These moraines are fed by the surrounding mountains. They are part of those mountains. Every day something is loosened by ice or frost or storm or avalanche, and comes tumbling down. Everywhere every mountain in the world is collapsing. Old mountains like the Wicklow Hills have by now been all but washed away; young mountains like the Alps are still formidable, but nothing can stay their ruin nor retard the wreck of their barren beauty – slowly but certainly the flakes of granite fall from their bastions like the petals of a rose. Consider the facts. The debris which falls into a glacier is subjected to a crushing and grinding power unparalleled by any other force in constant action. These mills of God grind the stones so exceedingly small that they are reduced to granite and slate dust, delivered by the stream issuing from the glacier in banks of black and white slime. The glacial waters from the chain of Mont Blanc carry away every year over 80,000 tons of mountain in the form of drifted sand and mud – to form at last fertile fields for human life. So every twenty years 1,600,000 tons come down – on that road alone, out of hundreds in the Alps. We stand upon the glacier and we think that it is steady – but we know that it is not. We look up at the hills and we think them everlasting – but we know that they are not. We may say that the life of a butterfly is short and that of a mountain long; but in the calendar of Nature the mountain indifferently shares the mortality of the moth and the worm. All those high places shall be brought low and all those crooked crags be made straight on the level plain where the waving grass is green and the harvest gathered.

And so, when we stand back and survey the scene, the whole scene of hundreds of glaciers and hundreds of mountains, shall we not say that those perishing peaks are – *seeds*. They are the seeds of future continents. Already from the summit of Mont

Blanc 10,000 to 12,000 feet of strata have been removed. The conglomerates of Central Switzerland, the gravels and sands of the Rhine and the Rhône, the Danube and the Po, the plains of Dobrudscha, of Lombardy, of southern France, of Belgium and Holland once formed the summits of the Swiss mountains. And shall we not say that the glaciers are – *ploughs*. And where is the ploughman? He is there all right, tirelessly at work. Let me use the words of John Tyndall, who is in the company of those few, in any age, who, laying hold of the facts, possess the added strength of imagination to bestow significance upon them:

And as I looked over the wondrous scene towards Mt Blanc, the Grand Combin, the Dent Blanche, the Weisshorn, the Dom, and the thousand lesser peaks which seemed to join in celebration of the risen day, I asked myself, as on previous occasions, How was this colossal work performed? Who chiselled these mighty and picturesque masses out of a mere protuberance of the earth? And the answer was at hand. Ever young, ever mighty – with the vigour of a thousand worlds still within him – the real sculptor was even then climbing up the western sky. It was he who raised aloft the waters which cut out the ravines; it was he who planted the glaciers on the mountain slopes, thus giving gravity a plough to open up the valleys: and it is he who acting through the ages, will finally lay low these mighty monuments, rolling them gradually seaward – 'sowing the seeds of continents to be' – so that the people of an older earth may see mould spread and corn wave over the hidden rocks which at this moment bear the weight of the Jungfrau[6].

4. Icebergs

Sometimes glaciers break off from the land and go to sea. The finishing touch is put to these extraordinary works when as ice-mountains, as islands of ice, they move across the oceans. Then we call them icebergs.

When we think of glaciers we must not confine our view to what can be seen in Switzerland. For a just estimate of what can

6. *Days of Idleness in the Alps.*

and does exist in this form, we must turn our gaze to the north, to Alaska where some glaciers are fifty miles long and three miles broad, and to Greenland which has been called the mother of glaciers; nor should we forget that terrible highway used by explorers to the South Pole, known as the Beardmore Glacier, 120 miles long and with an average width of twenty miles.

The enormous ice-sheets of Greenland and Antarctica covering 500,000 and 3,500,000 square miles respectively, and probably from 8,000 to 10,000 feet thick, are partly enclosed by mountainous rims through which the ice escapes as valley glaciers. Much of the Antarctic ice-sheet, however, is unconfined in this way; the ice overflows the coasts, and presses out to sea as colossal floating ice-barriers (e.g. the Great Ross Barrier) which terminate in high cliffs from which the great tabular icebergs of the Antarctic seas break off.[7]

That is the chief region from which they come, though by no means the only one. They leave many shores. They are met with far out at sea. Pale promontories, broken from what bending cliffs? Is this the ghost of a mountain that we see approaching in the moonlight? Mariners must have asked such questions faced with such shapes. They have likened them to many things as they have gazed upon ruined castles and windowed walls, upon abandoned citadels with dismantled towers and soaring columns and graceful arches and massive obelisks, as also upon brutal chunks and bare escarpments exiled from the land. Some have attained extravagant dimensions. They have been known to measure from thirty to forty miles in length and twelve in breadth, rising to 1,500 feet out of the water; and on 17 January 1893 the vessel *Loch Torridon* came upon an island of ice which measured fifty miles in one direction, and having sailed to the end of that side, no one on the ship could discern, even from aloft, any termination in the other direction. These giants are more often called by whalers ice-fields. They sometimes take the shape of mountain chains such as might be found on a

7. *The Earth and its Mysteries*, G. W. Tyrell.

large island. 'A deep layer of snow completely covers the floating fields, and one might imagine them to be Swiss cantons wrested intact from their foundations in mid-winter and cast into the sea, where some mysterious force keeps them afloat, with their plains, their ranges of hills, and their valleys.'[8] Are these travelling countries inhabited, one is tempted to ask. Yes, polar bears have often been discerned in residence, embarked for unknown shores, while Greenland ice-packs have not seldom provided accommodation for travellers whose ship has become stuck in the floes, like that group of nineteen people, including women and children, who in 1871 were forced to quit the American ship Polaris, and sailed away on an ice-pack for seven months on a journey of 1,700 miles.[9]

Icebergs are launched singly, of course, and they are often encountered singly – but not always. Sometimes they are seen advancing in fleets extending as far as the eye can see, 'dancing up and down on the waves' in Fabre's striking image 'like the ruins of some city of giants, built of crystal, alabaster, and marble.' They were once the terror of all sailing craft in those regions. Who could face such an Armada! The colossal masses swayed about, often closing up and crashing against one another, and woe to the ship that was caught between them. It was not always easy to escape, and Scoresby, the English navigator, relates how in a single summer no less than thirty vessels were crushed like eggshells (frail as the tiny *Titanic*), sometimes leaving nothing visible save the tips of their mainmasts.

So these icebergs sail the seas. As glaciers their pace was slow. In their new element, and in their altered guise as ships of ice, they gather speed. They do not vaguely drift, they journey forward firmly. Where then their rudder? What their sails? And who the captain? They are drawn ever onwards from the Arctic to the Equator in the arms of the great *river in the sea* called the Polar Stream. They have been known to travel 2,000 miles

8. J. H. Fabre, *This Earth of Ours*.
9. See W. Coles-Finch, *Water, Its Origin and Use*.

before disappearing. Did we doubt their size, this should prove it, for during all that time their fainting shoulders and their steaming towers paid back their tribute to the sun.

They must needs pay that tribute. For these leviathans of the sea do not belong to the sea, they have crawled forth from the land, they have come down from the sky. Yet they do belong. They do belong, and have come at last to their own rightful home again. They started here; they rose on high; they fell; they flowed; and now are back once more. Drastic symbol of all our fleeting earthly shapes: today visible, tomorrow invisible; today a liquid, tomorrow a gas; today a solid soaring vessel riding on the waves, tomorrow as the flying vapour.

5. Frost

If snow is created in the heavens above and ice on the waters below, what is this other thing that we find after the freezing night clinging to the branches and the twigs, that is not snow and is not ice? It looks like snow, but is finer and less weighty – and, on inspection, is it not ice? It is an aspect of both: we call it frost – more properly hoar (or white) frost, as opposed to frost which also bites but does not decorate. It can be described, as far as I can see, as snow which has not been formed above in the cloud and then come down in a conglomerate of feathered forms, but has taken as its foundation not the dust of the air up above but solids down below. The twigs have caught the freezing, foggy vapour as in a net and solidified it. Frost is made of snow crystals that have not fallen; it is earthly rather than heavenly snow. Thus if we know what has gone to the creation of ice and snow we also know the essentials in the making of frost.

This drapery brings with it no spectacular tale of human woe such as we get through snow-caused avalanche and flood. Men's lives are not endangered, but it is the terror of fruit farmers, since it may come just when the ascending sap and peeping buds of fruit trees need warmth. I have just said that it is really the same as snow: but because the crystals have used the trees for their building base they freeze the juices, while snow, already based, softly settles upon the trees and bushes, doing no more damage than cotton-wool. Hence the fruit-growing department of agriculture calls for a farmer of stout heart. As I have put it elsewhere –

He who deals with arable land can count upon a certain degree of harvest in the worst of seasons. He sows his seed and according to his skill in husbandry he will reap his reward. Luck, sheer luck, may elevate or destroy the fruit-grower. The beauty of spring, the whole parade of bloom and blossom, can change overnight into the whiteness of the flowers of frost. The spectacle of promise and bounty turns into a picture of blasted hope.[10]

Apart from the unfortunate men upon whom we rely so much for the fruitfulness of the table, the rest of mankind can afford to sit back and gaze upon a transfigured world. It is no less. I lived once in a house in Kent with trees and bushes just outside the sitting-room window which received the morning sun. After one very cold December night I went into the room and was startled by the transfiguration of the trees and shrubs outside. Under the rays of the sunshine I saw no trees and no bushes, but an emblazoned armoury glittering with countless facets. Not all the cutlery in Sheffield, not the Crown Jewels, not the fabled diamonds and rubies of princesses in the palaces of Eastern kings could have competed with the show outside made with water and air. At another place, in Dorset, by taking a few steps from my cottage, I commanded a view of a birch-treed vale. One morning, after a similar frost, I looked across the little valley and for a moment thought, really thought, that the trees opposite were in a white cloud. We have noticed how clouds do sometimes form in the valleys – as fog or mist – when trees and housetops are seen rising from them. But on this occasion every particle of every branch and twig was cloud-clothed. I speak with exactitude, for at first I thought it was vapour. It was arrested vapour; frozen, solidified, stationed,

10. *While Following the Plough*

frosted – bricked by myriads of crystals, each beauty's gem and art's despair. It reminded me how when once gazing at the trees standing thickly on the hill that answers the great Schaffhausen Fall I had been bewitched by the lacery of that furry caterpillared coat of stainless white on all the branches and on every tiny twig; and yet so many numbered that to the distant eye a mist was made within the wood as in reply to the mist around the fall.

Yet not until we come close can we really take in the full wonder of this work. All things are transfigured, even barbed wire and iron railings. Nor does it matter how tender or small or intricate the object is; without smudging and without weighing down or breaking, the white print is cast. The spider's web hangs on the gate. Yesterday we might not have noticed it. Today it has caught and fixed the freezing vapour in its net – a photograph developed from the negative. We behold a double miracle: the weaving of the web by the living, crawling creature, and the weaving of the filaments of frost by the molecules that in another mode tread another loom. And when we draw near and examine the frosted figures formed and forming on the window-pane, we enter fairyland. The ghostly fronds and silver forests in the starry night, the castles and the cities, the birds above the tropic trees, the corals by sad abandoned shores, lead and lure us on into a land we feel we know and yet have lost.

How brief these exhibitions are! There was a good frost last night, and in the morning I went out into the garden and had a close look at the display. The rose trees were in flower again. On every branch, on every stem, on every thorn the pearly blossoms bloomed. Night and called them: darkness was their cradle, the freezing sky their swaddling. I looked round the garden, where every bush and fruit tree blazed with these white garlands; and I thought of how the trees had shone in spring with apple, may, and cherry blossom, and soon would shine again when sun and soil had done their work for them. Rootless, these present flowers had taken nothing from the soil; and, leafless, nothing from the sun. It was a clear blue-sky morning.

The sun increased its circle. As it climbed they slipped and fell. As it grew in strength they waned and withered. And when it reached its zenith all those gay blossoms had perished from the earth.

CHAPTER II

SECTIONS FROM
INSIDE THE MOUNTAINS

1. The Raindrop

It rains. When and why did the first drop fall upon the earth? It will be as well to get this clear before we proceed. At first there was only one thing in evidence – fire. There was no earth. There was no water. There was no atmosphere. More strictly, they were present, but not evident: they were within the first – but there was no distinct lithosphere, hydrosphere and atmosphere. The fire cooled (this cooling of heat being as wonderful a mystery as the getting hot of heat, but not to be considered here). The fire cooled into molten liquid and then into solid rock on the outside, a folded and faulted crust – and this was the beginning of the lithosphere. At the same time it threw off a vast congregation of vapours which included all the water that later was to be present in the oceans and the seas and the lakes and the rivers – and this was the beginning of the atmosphere. Thus there was now a lithosphere and an atmosphere, but the oceans were suspended in the air above the land. Then they fell down upon the land – and this was the beginning of the hydrosphere.

It was not easy for the water to establish itself below. The rain came down in torrents, in vertical rivers, literally in oceans – and the bare and burning rocks threw it back again. A battle raged between fire and water for a long time, no one knows how long, nor may the mind easily conceive those embattled scenes between vapour and vulcanicity raging through endless eons. Finally, the victory went to water. It enveloped the *whole earth*. Then nothing could have been seen under the sky except the shoreless waves. In the first movement, when fire was triumphant, there had been a lithosphere and an atmosphere, but no hydrosphere: in the second movement, when water was triumphant, there was an atmosphere and a hydrosphere, but no lithosphere – in sight. But this second movement was no more permanent than the first. Just as the water could not be held above in the sky, so the fire could not be held below in the earth. It burst forth, rending the rocks into deeper basins and higher hills. Then that huge single ocean flowed into these new depths, with the result that the water under the heaven, as Genesis has it, were gathered together into one place and the dry land appeared. Because the fire was unable to resist the water and because the water was unable to quench the fire, we got the oceans and the continents – though the distribution was unequal, dry land being only one-third in the proportion. Moreover, such dry land as did appear at first was very far from having the extent of our present continents, and there were to be many changes and readjustments, many risings and sinkings, before we got the present configuration of the globe.

There they were, then, the oceans and the continents: two separate spheres. But this did not mean that they were to be wholly separated. The division was made, but the oceans continued to flow over the land in modified form. This is called raining, and we have studied the means by which the atmosphere accomplishes this. It is what I would call the fourth great movement: first, fire; then water; then land and water; and, finally, the dry land not left dry, but watered by the ocean in the form of rain. Now let us see what happens to the earth and on the earth in the process of this watering.

The rain falls upon the earth – upon the mountain, the hill, the slope, the field, the plain. And if we ask broadly, what does it do and where does it go? we know that it seeks the sea, it immediately moves towards the sea; if it meets with a hollow, it will fill it and make a lake and then proceed; if with a mountain it will go round it; if with a cliff it will go over it – always seeking and eventually finding the sea. So much is common knowledge, evident to all of us. Yet all of us also, in our childhood, and perhaps long after we have grown up, have asked ourselves in wonderment as we stood beside a river; How *can* this volume keep up its flow? I confess that I always used to wonder about this, whether I stood beside a stream, or a river like the Danube, or a waterfall like the Schaffausen. How can it keep it up? Surely by this evening there won't be any water left at the source.

Childlike questions of that sort are seldom really foolish, and I do not think that there is much wrong with this one, for the rivers certainly would not keep up their steady flow if the rain merely fell upon the surface of the earth and ran along it. Rivers are recruited by reservoirs beneath the surface. In some cases the term 'reservoir' must be used figuratively as meaning seepage – that is to say, the river rises from a huge, saturated, trickling sponge which is always being replenished from the vast reservoirs of the sky. That is one way by which rivers are started and fed. It implies that rain begins by entering the earth rather than running along its surface. This is, of course, quite obvious, and all of us realize in rather a vague sort of way that there is this seepage and that many hills must be sponges which, receiving the rainfall, absorb it and distribute it – though still the constant flow of a great river seems a marvel. But it is only in the last decade that we have become at all well informed about something much more interesting – namely, the underground reservoirs and lakes and waterways in certain mountains, especially those with limestone properties, which also feed the rivers and springs and wells of the world. For all who are moved by the poetry of geology the underground realms carved out by water make a particular appeal to the imagination. A study of water opens up many vistas; we pass through many doors and see much that is unexpected; we come now into strange lands indeed, the country of the speleologists, the explorers of the underworld.

A general preliminary statement of the facts will at once hold attention, I think. We approach a mountain range – the Pyrenees, for example. It looks solid enough, that pile. But it is not solid. We can go inside. We can take a walk in the bowels of the mountain. We can enter at one end and come out at the other side. We can boat or skate on lakes within it; we can go down shafts for two thousand feet; we can pass through corridors and galleries leading to magnificent halls; we can journey beside waterfalls and cascades, streams and tempestuous rivers; we can gaze up at cliffs that pass out of sight, and down over terrible precipices; and as we pass along we will behold from time to time the filigrees of flora from the mint of minerals. Such are some of the facts concerning our journey in the mountain, one of many possible journeys in mountains – though few as against those undiscovered and those which can never be found, since they have no outlet.

How has all this come about? Descending one of those shafts, standing in one of those halls, entering into one of those corridors hung with stalactites and pillared with stalagmites and gemmed with the flowers of gypsum, we might well ask: Who sank this mine? What architect built this temple in which the temples of men could be housed? What artist hung these draperies? What soil and what sun brought forth these flowers? Where the chisel and whose the hammer for the excavation of these caves?

And we must answer: The raindrop.

It is always thus with Nature. Wielding the tool of eternity, she works by softness and insinuation. In her hands the humble are easily exalted and the proud rebuked, the weak vessel acting with astounding power against the strongest forms, the fluid drop holding in itself the force to wreck the hardest rocks.

The rain falls upon the mountain. Let us cast our minds back again across the centuries, across centuries of centuries. Let us take it that the first rain is falling upon the Pyrenees at the beginning of their term. The rocks will contain many cracks and fissures, many crevices, clefts, and cups. The rain enters into these places. Some of it overflows and runs down the mountain, some evaporates and returns to the sky. What happens to the remainder? It acts. Raindrops are not passive, but active agents. They have been in contact with the atmospheric gases and have come down informed with their properties – especially with carbon dioxide. This preparation eats rock. By process of what we call erosion and corrosion, with endless time to do it in, the raindrops enter into the mountains and begin to carve out cavities within them. Raindrops act as knives and chisels; and when in their accumulation they become torrents, then they act as hammers knocking away what the chemicals have already loosened. Grottoes and caves occur in every kind of rock, in lavas, basalt, slate, and granite as well as in limestone, dolomite, and gypsum; but, owing to their particular fragility and solubility, calcareous rocks yield most easily to the process of being undermined, and in the course of ages whole mountains may be hollowed out.

In the nineteenth century various mountain caves were known, in Austria, in Australia, in France, in England, and in America, where the seventy-mile accumulation of avenues which compose the Kentucky Mammoth Cave is still considered according to Martel, to be 'one of the most impressive phenomena to be seen on this planet of ours.' But it is only recently that men began to realize that here was a new world to be conquered. It had seemed that the earth now offered nothing to explorers; that the virgin forest, the untrodden desert, the mapless mountain, the unjourneyed sea were all things of the past and that there were no realms remaining which were completely undiscovered. We were tempted to say, 'It is finished: there is nothing left now for the iron-souled adventurers, the active visionaries, who in all ages have sought to penetrate into uncharted places, and with nothing more than an intuition to go by, and with the possibility of irretrievable calamity facing them at every step, have yet endured all things, overcome all things, and discovered new realms. Now there is only the moon left.' Yet we need not say this. There is speleology. We can still go where others have not dared to tread or have not found the way. We can enter regions still waiting to be explored by brave men. We can open gates hitherto closed and pass into territories of inconceivable beauty and alarm. We can find chambers once known before the dawn of history by cave men, and then penetrate beyond into places which have never previously echoed to the trampling of any feet. This is the country of the speleologists, this their playground; here lies all their ambition, their desperate endeavour. They are not as other men: unexampled fortitude, steel-like patience, sublime audacity, the nerves of acrobats and the blood of fishes, belong to the adventurers whose way of life is cast in those lonely lands.

2. The Adventures of the Speleologists

The French speleologists have been the most articulate, and it is to them that we chiefly owe our knowledge, and through them we may experience marvellous and terrifying adventures by proxy. The honoured pioneer is Edouard Martel, who led the way at the close of the last century and the beginning of this one. 'Here was a man,' says Norbert Casteret, his generous-minded follower,

who descended to terrifying depths to look at underground France, and who discovered the Palace of the Thousand and One Nights for the greater glory of science. He thought nothing of risking his life daily by having himself let down on a rope tied to the middle of a stick; he rode his broomstick clad in a bowler hat and a sack coat like a Jules Verne scientist. Martel was 'the man of the chasms' ... He has explored hundreds in many countries.

Now the number of speleologists increases yearly and tales come in not only from France but from England, Italy, Switzerland, Austria, Algeria, Yugoslavia, Montenegro, and the United States. Pierre Chevalier, who has spent 960 hours underground and gone on fifty-nine out of the sixty-five of the expeditions carried out by the Speleo-Alpine Club between 1935 and 1947, lists forty cave systems having a total drop of over eight hundred feet – though some go down more than twice that depth. And these underground systems, as I have mentioned, are but few in comparison with those yet to be discovered, and are nothing in numbers when we think of the systems which offer no means of approach.

What are the means of approach? Roughly two: either you

see a stream disappearing into a cavity in the mountain, or merely a hole – perhaps not very obvious save to the trained person who is looking for it. They may be too small to offer any entrance to a man, or just large enough, or extremely large at the outset. This is where the underground mountaineer starts his adventure.

Often he starts alone. Later on, if he has found something interesting opening up, he will seek company; later still a team may prove necessary. I am thinking especially of Norbert Casteret. He was born with a genius for speleology. Caves called to him at the age of five. From then onwards they reigned supreme in the empire of his thoughts. He opens one of his books with the words: 'I know and love caverns, abysses, and subterranean rivers. Studying and exploring them has been my passion for years. Where can one find such excitement, see such strange sights, enjoy such intellectual satisfactions as in exploration below ground?' In recounting his search for the true source of the Garonne (an important geographical and political point) he drops the remark: 'For three years I had not passed a day without asking myself, "Where does that stream go?"' That is typical of the truly dedicated man – the counterpart of which question could easily be quoted from the utterances of all such men in many different fields, especially in science and art.

Casteret loved exploring alone. In fact, in the course of his cave career he made the most dramatic solo adventure in the annals of speleology. He came to a crack in a mountain at Montespan in the Pyrenees, from which a stream was running out. He undressed, squeezed through the hole and found himself in a corridor about 12 feet wide and 8 feet high. Wading in the water, he progressed about 60 yards when the ceiling began to dip, the water to deepen, after which presently the roof and water joined. Thus he was stopped.

Explore farther? When men, as opposed to fishes, come upon the roof of a cave dipping down into the water, they can go no farther. Yes, but your passionate speleologist is not as other men are. Why not dive into this depth and this darkness, Casteret asked himself, and see if it is an underwater tunnel through which one could swim and perhaps come up again into the air at the other end? True, the risk was appalling to contemplate: the tunnel might go on indefinitely, he might go over a high waterfall, he might run into a cul-de-sac, and if tempted too far might fail to get back before his lungs burst. Almost up to his neck now in water, with one hand clutching a candle, Casteret stood there in the awful loneliness and silence weighing his chances. Then sticking his candle on to a ledge, with a mixture of confident intuition and sublime audacity he dived into the unknown, using his finger-tips for eyes – *and came up the other side.*

He had forced what they call a siphon, a tunnel with a submerged ceiling. Then with a good sense only equalled by his courage he immediately turned about and dived back again – for to lose sense of direction in that total darkness would be as easy as it would be fatal. So back he went to his candle shining on the black water.

Next day he returned with more equipment, consisting of a rubber bathing-cap full of matches and extra candles. Once more, clothed only with these things, he entered the freezing water in the dark, cold underworld, and again successfully swam through the siphon. His candle now showed him a roof parallel to the water with a thin air-space between. After a hundred yards he came to a clay bank at the entrance to a chamber rising to a height of forty feet and decorated with stalagmite cascades. Accustomed as he was to caverns, he confesses that he had never known such a feeling of 'isolation, oppression, and terror', – for these men do not possess less sensitivity than others, only more courage. He found himself now faced with another siphon. He dived again. It was longer than the first, but again he got through. Now he was locked below by a double barrier, and in the untellable solitude and black silence he 'struggled against an uneasiness slowly turning to anguish.' Shaking with cold, numbed by the glacial water, he yet pursued his way; crawling on for some distance, he reached a larger hall than the last. He continued his exploration for five hours through further chambers and corridors and galleries with ever more perspectives opening up and leading him on whenever he thought he had come to some impassable bottle-neck. At last he went back, nearly losing his way when confronted with forks he had not noticed on his way in, and failing to get through the worst siphon at the first attempt.

A year later he returned to the attack, this time accompanied by a friend. Having once more passed the second siphon, Casteret began to probe in a nook with an excavating tool he had brought with him on this occasion, and came upon a primitive flint which had obviously been used. No sooner had he discovered this than he came upon a clay statue of a bear which he had not noticed in the feeble candlelight.

He had stumbled upon a piece of sculpture that was soon to be acknowledged by scientists from many countries as the oldest statue in the world.

Then for an hour discovery followed discovery – horses in relief, clay lions, engravings, designs, mysterious symbols. All the important large animals of the Magdalenian period were engraved on the walls and rocks. Many of them showed mutilations – as if javelins had been hurled at them. Sometimes they found claw-marks beside human footprints in the clay, and sometimes both were mixed together where there had been a struggle for possession of the cave. A footprint in a lonely place is always a pathetic and a tragic sight. Here were footprints, forlorn and fearful, that had remained in solitude for two hundred centuries of speechless night.

Thus was the courage of Casteret rewarded. He had plunged into the unknown – and had come up into the Past to roam in ages before the dawn of history.

It is no doubt a commonplace of archaeological adventure to come upon cave drawings – though never less than thrilling – and to contemplate the combination of art and magic. We started in caves, and at the death of civilization the other day, when the three necessities of man became food, clothing, and shelters, we returned to them – and may yet be compelled to do so on a much larger scale. In between these extremes caves have provided habitation for hermits and other lovers of solitude throughout the ages, for many saints in many countries, for professional wise men, for beggers, and for those of whom it was written that they 'went about in sheepskins, in goatskins: being destitute, afflicted, evil entreated (of whom the world was not worthy), wandering in deserts and mountains and caves and the holes of the earth.' There is nothing exceptional in discovering the relics of habitation in caves, but it must be conceded to Casteret that the way to his discovery no less than what he found provides the most dramatic instance in the history of such enterprise. It is with some amusement that one learns how, when the distinguished scientists arrived to check his claims, they were baulked by the siphons, and the water had to be considerably drained and diverted before they could attend at the primitive exhibition.

That adventure and triumph of Casteret's is moving and thrilling, but perhaps one is even more moved when the explorer pursues his ways and comes, after frightful hardship and danger, into regions which hitherto no man had ever penetrated since the foundation of the world – that surely is the ultimate of exploration. To emerge from a cat-hole into the soundless blackness of eternal night, to stand alone in the unspeakable isolation of a sepulchral hall – imagine the anguish and the ecstasy!

A cat-hole is horrifying to contemplate. It is a hole which the explorer thinks he may possibly be able to squeeze his way through if he lies on his stomach and worms his way along. He cannot tell what length the hole is, or whether he may fail to get through, and perhaps stick. It has happened more than once that a speleologist, on his own, miles from all human succour, has got stuck, neither able to advance nor retreat, and there has perished. Yet nothing seems to daunt these men, and they have frequently unlocked their door by climbing through the keyhole. Also, entrance to this underworld often must be effected by descending grim perpendicular holes for about 800 feet. These pot-holes are nearly as frightening as the cat-holes. One of the bravest of these adventurers, after gazing down into his first pit, declared afterwards (though too modestly): 'I realized with shame and despair that I was no explorer. I would never dare attempt such a chasm – the very sight of it scared me to death.' Worse in a way are the journeys down the waterfalls. After descending a pot-hole of 500 to 800 feet, perhaps, and finding a landing-place at last, he may thence make his way through a hall or a corridor – the possibilities are many. He may be compelled to wade through a lake and along a river which suddenly falls over into another shaft. Is this the end of his journey? By no means. If he has enough rope with him he will tie it to a rock and go down with the waterfall. If the rope is not long enough and he fails to reach the bottom, though already soaking and frozen, he will be obliged to climb up against the cruel buffeting of the torrent.

Worse possibly is their special method of being hanged. When a party of speleologists are tackling some formidable cave they sometimes place a windlass at the top of one of those terrible holes which confront them as they proceed. By this means a man can be lowered by cable down into the darkness. The cable is 1/5 inch thick. Its composition is such that it is very strong, but its thinness strikes terror into the toughest of these men, and they always feel that they are literally hanging by a thread. The man is lowered, sometimes hundreds of feet, until some landing-place is established where he can be joined by a comrade, and together they will continue their journey into the bowels of the mountain. But in due course it is necessary to get back and be hauled up. On these occasions the windlass does not always work. It frequently breaks down, gets stuck, advances upwards in jerks, or suddenly lets the man down a few feet while he hangs there in agony of expectation. Casteret gives an account of how his companion, Vander Elst, thus ascended from a pit in the Basque country. It was a question of stoppage after stoppage, a succession of jerks, prolonged halts – repeated for nearly an hour as he hung there. Casteret speaks of his anxiety for his friend as he watched from below, seeing him swing slowly higher, when the breaking of the cable would be fatal, and how if the winding apparatus broke down definitely he would be suspended there indefinitely without any possibility of rescue; and he adds these words (without emphasis): 'Not for one moment did he show the slightest sign of being worried, *nor did he address a single question to the men above*, who must have been having an awful time with the winding, coupled with the dread of not being able to get us up.'

They are rewarded for these labours. The raindrop has seen to that. The landscape which it has opened up for the travellers in this far country exceeds their expectations and justifies their toil. They are amazed at the revelations resulting from the hydrological ablutions carried out through centuries of dawnless dark in the depths of the earth. The explorer never knows where his steps will take him, what he may encounter next. At the other end of this siphon or that cat-hole, at the bottom of this shaft, what will declare itself? Will it lead him to a narrow corridor or a cathedral nave, to a river or a waterfall, to a lovely sea-green lake or another yawning pit?

A huge vault may open before him, a chamber so vast that, like the giant cave near Trieste, the massive St Peter's in Rome could be put inside it, or like the cavern of Cagire, into which a football field would fit, together with the Cathedral of Notre Dame with its thirty-seven chapels; or the chamber itself may be like a cathedral, and it was right for Casteret and his companions to celebrate Mass in the Grotto of Esperros with its Gothic nave, its pillars of stalagmite, its altar, shrine, and font.

He may discover at the bottom of a chasm a grim chaos of boulders and all about him the results of explosive pressure and primitive savagery of convulsion; and yet just beyond this he is likely to come to a miniature grotto bearing witness to the delicate, graceful work of water falling drop by drop to create slender, translucent stalactites, filigrees of gypsum, and embossed draperies upon the many-coloured walls; and roaming farther he may find himself gazing across at a dream country with stalagmitic mosques, minarets, and towers on a rounded white and yellow hill, belonging to the *Arabian Nights* realm of the mind; while at his feet is a limpid lakelet with water so clear and quiet that he can see the bottom as sharply as through glass, a submarine landscape starred with crystalline fronds like a coral carpet – perhaps the source from which Alph the sacred river ran.

He may pass into a void, or what seems to him a void, and wrestle with impenetrable nothingness, his candle revealing only a hall of empty darkness. Alone he has entered into a loneliness unmeasured; a solitary in a solitude like no solitude upon the earth, he is suspended in centuries of accumulated silence, broken only by the steady fall of a single water-drop like the ticking of the clock of eternity. He may stumble on into a deafening region of roaring rivers and sounding cataracts; and then come to another place where the waters have been stopped and stilled: the waterfall frozen into silence and chiselled into peace; an ice curtain hanging free; the lake a mirror of glass; the river like a glacier; and huge towers rising in the abyss like icy arms upraised to stay the further steps of human feet.

He may come to a spot where he hears and is sure that he hears voices, where no voices are; where he is positive that a conversation is going on, for he can all but catch the words, though there is no one to converse; where sighs and wailings come to him from places where there is no one to lament. If he reaches a cavern above which a gleam of light shows that there is a hole leading to the surface, he may stumble upon a charnel-house of decaying corpses of animals that have fallen down. Sometimes mouldy wood is also found in such a spot, and the astonished speleologist may encounter phosphorescent mushrooms gleaming in the darkness like glow-worms, such as were seen in a pot-hole in the Basque country – a vegetation as extraordinary as the nightmare meadow of thick white grass that graced the bottom of a chasm, three hundred and twenty-six feet down in the Moroccan Atlas.

He will not often see such monstrous shapes of tragic growth in sunless soil; his gaze will be turned far more often to the true plants that furnish these regions with appointments more resplendent than in the palaces of kings. The stalagmites rising from the ground, the stalactites hanging from the roof, are the most enthralling of all the marvels of water-work. They grow at the rate of about an inch in 1,000 years, or perhaps the thickness of a sixpence in the compass of a man's lifetime. Often they meet; but not in the middle, not halfway, for it is obvious (or so it seems to me) that the stalagmites build up a good deal faster than the stalactites build down. When they meet they look like pillars supporting their temple. A small *forest* of them, numbering 400, from 60 to 100 feet high, were found in the Chasm of Armand in Lozère. Gazing upon them, the stupefied explorers felt that all their toils and dangers were as little against the reward of such a revelation of Nature's engineering: first the raising of vapour by virtue of the sun, and its housing in the lofts of the sky; thence its descent upon the mountain and into the rocks, followed by the excavation of the chasm; and, finally the tribute of the pale deposit which through further centuries had built up these silver trunks of stone.

We generally think of stalactites as daggers and spears and Damoclean swords, but they assume all sorts of shapes and all sorts of colours, sometimes looking exactly like bunches of carrots (pink and yellow), or so like curtains that you feel you could easily take them down and fold them over your arm. We need not go far to see these things, but we must go a long way, we must climb up to an opening in a cliff, and then crawl through a cat-hole, to see what shows itself in the deepest abyss in France, the Cirque de Lez. The first man who got through the cat-hole shouted frantically back to his comrades, '*The white sea!*' Scrambling through after him, they beheld a brilliantly white surface spread under their feet, terminated by high white cliffs – none of it made of water or ice, but stalagmite. The white expanse narrowed and became like a fjord between those Dover cliffs, until it ended at the foot of a petrified waterfall. They continued their journey, encountering fresh marvels which their most experienced leader was quite unprepared for. Advancing through chambers and vestibules now, unable to avoid trampling on flowers, crushing masses of crystal with their hobnailed boots, breaking glass rods and swords and coral-bushes, they stepped into a fairy palace. 'Stalactites and crystals sparkled everywhere; their profusion, their whiteness, their shapes were fantastic beyond belief. We were inside a precious stone; it was a palace of crystal.' As they walked on they brushed against strange cobwebs that broke at a breath, and against 'silver strings with the brilliance of silk yarn', dangling from roof and walls, which could be tied in knots and wound into balls. These mineral cords and cobwebs were a form of gypsum, which provides the most varied floral display in this kingdom, in crosier and plume and cluster and gem. Was this the thread of which Blake spoke, which if you held on to would lead you in at Heaven's gate?

We must now leave these scenes. It has been encouraging to meet these men. As we go from their presence many images come before the mind. We think of them passing from the sight of peaceful forests, down the pitiless shaft into a world totally and endlessly dark, to wage their battle and match their skill against indefinite difficulties and obstacles, pushing their way ever farther into the heart of a mountain. We think of their special and dreadful hazards: getting lost in a labyrinth, or stuck in a cat-hole, or drowned by a sudden inundation, or injured by falling stones, or nearly frozen to death in drenching clothes. We think of the patience of men who can perch on ledges or narrow balconies 400 feet down a chasm overhanging pitch-black space for fifteen to twenty hours while their comrades in front carry on the exploration, hearing nothing save the shrill little cries of bats, sad children of eternal night. We think of the perseverance that drove them to return again and again to the attack of some desperate, ever-opening cave such as the Henne-Morte (after the dead woman whose winding-sheet was snow and her tomb a mountain), which was attacked eleven times during ten years before yielding to conquest.

It was during this saga of assault that two men were terribly injured in the depths, one by slipping, the other by a stone falling on him, and had to be got back to the surface – a terrible and wonderful story. When thinking of those two men, whose one desire as they lay in agony was to be able to climb and search again, my mind turned to another scene: I thought of Philip Wills's glider-flight over the Italian mountains and of his dying comrade who had met with disaster, declaring that he would yet live to fly again. I thought of the two lots of men: the glider pilots who, mounting through the shafts and potholes in the air, hold acquaintance with the clouds, and stride the blast, and range the kingdoms of the sky: and these their fellows who leave the light of heaven for the darkness of the pit to search the secrets of the caves, and stand beside the Styx in the empty hall of Hades.

3. The Scenes of Petrification

Hitherto . . . I have spoken of the raindrop as a solvent carving out chambers and avenues in the earth. True, it takes away. But it also gives. It gives less than it takes away, and, strictly, it adds nothing to the whole. Yet when it has taken from one place it may give to another place. That which it has hollowed out of one rock it may put into the hole of another rock. It may chisel out a crevice, but it may fill another crevice. This is known as deposition. Water can close up cracks by depositing minerals. 'If in the chemical laboratory,' says Professor Salisbury in a succinct statement, 'solutions of various sorts are mixed in a test-tube, some of the materials in solution are likely to be precipitated. The same thing takes place in the rocks beneath the surface.' The mingling of waters coming from different directions may effect a chemical change causing some of the material to come out of solution and be deposited in the crack. And when cracks are thus filled up with mineral matter the rocks are said to possess veins. In such veins the most valuable ores are sometimes deposited – and the deposits in these banks are often of gold and silver. Instead of these fleeting forms being lost for ever in the river or cast loose upon the plain, they are gathered into the crevice and harboured in the hole. For the raindrop not only erodes the earth but assists at the creation of the precious stones that dazzle mankind.

The most spectacular examples of beautiful deposition are the stalactites and stalagmites which we have already noticed.

But they cannot compete with Hierapolis. In that portion of his mighty work called *Attis, Adonis, Osiris*, Frazer speaks of the chasm of Hierapolis in the upper valley of the Meander on the borders of Lydia and Phrygia, from where hot springs rose, the vapour of which caused instant death; and he tells of the ceremonies carried out there when a company of naked youths, their bodies glistening with oil, would carry up to the mouth of the chasm, a sacrificial bull which on breathing the fumes expired on the spot. 'But another marvel of the Sacred City,' he continues,

remains to this day. The hot springs with their calcareous deposit, which, like a wizard's wand, turns all that it touches to stone, excited wonder of the ancients, and the course of ages has only enhanced the fantastic splendour of the transformation scene. The stately ruins of Hierapolis occupy a broad shelf or terrace on the mountainside commanding distant views of extraordinary beauty and grandeur, from the dark precipices and dazzling snows of Mount Cadmus away to the burnt summits of Phrygia, fading in rosy tints into the blue of the sky. Hills, broken by wooded ravines, rise behind the city. In front the terrace falls away in cliffs three hundred feet high into the desolate treeless valley of the Lycus. Over the face of these cliffs the hot streams have poured or trickled for thousands of years, encrusting them with a pearly white substance like salt or driven snow. The appearance of the whole is as if a river, some two miles broad, had suddenly arrested in the act of falling over a great cliff and transformed into white marble. It is a petrified Niagara. The illusion is strongest in winter or in cool summer mornings when the mist from the hot springs hangs in the air, like a veil of spray resting on the foam of a waterfall. A closer inspection of the white cliff, which attracts the traveller's attention at a distance of twenty miles, only adds to its beauty and changes one illusion for another. For now it seems to be a glacier, its long pendent stalactites looking like icicles, and the snowy whiteness of its smooth expanse being tinged here and there with delicate hues of blue, rose, and green, all the colours of the rainbow. These petrified cascades of Hierapolis are among the wonders of the world.

The hot springs rise in the midst of the ancient city, and their deposits have raised the surface by many feet, and the white ridges conceal much of the vast and imposing ruins. In antiquity the husbandmen, by encouraging the water to flow in rills round their lands, found that in a few years their fields and vineyards were enclosed with walls of solid stone.

Erosion and deposition may sometimes be going on at the same time and even in the same place. Water may be dissolving a substance and at the same time be depositing another substance. While a rock is being robbed of its properties it may be receiving other properties. One rock may even become another rock. This is known as petrification. The object that suffers metamorphosis need not be rock in the first instance. Thus the substance of a shell, or a coral, may be changed, though its form is preserved. Wood can be petrified. A forest may be changed from wood to stone, and the astonished traveller, such as Antoine de Saint-Exupéry in the desert between Cairo and Alexandria, may find buried in the sand tree-trunks and boughs all turned into solid marble. The life and death of that lost forest had long since been washed from the memory of man, but an effort had been made to preserve it in the more lasting memorial of marble.

The dying poet spoke despairingly of one whose name was writ in water. But it is through the agency of water that names are written and tales unfolded reaching back into the gulf of time. What else forms the fossil but the flowing flint? The mineral in solution met and remained upon the body of the living form: the body and the bones perished, but the creature is with us still, for the deposit clothed the figure and sketched its shape for ever, and we read its history as in a book. Thus there is the carving out by water, the subtraction; and there is the addition, the preservation, the history work, so that the first and lowliest of all the creatures of the earth are furnished with tablets that will outlast the time-defiant monuments of the Pharaohs.

CHAPTER III

SECTIONS FROM *THE WORK OF RIVERS*

1. The Arrival of Plants

We must now come to the surface. We have witnessed the work of water underground. We have seen how it is held in reserve and perennially emerges in the form of springs, and can be tapped by wells.[1] We have entered the territories which it has created in the mountains. But it is a cold and flowerless land in which nothing living can be nourished save a few blind insects in the desolate pools; no fish could live in those lightless lakes, as no bird could sing in those shades – only the poor little bats, shrinking and fearful, might find a home. How different it is above ground. There we see that water, accompanied by light and warmth, is the giver of life.

At the beginning of the last chapter I recalled the time when the earth consisted only of fire and of rock. The main fact to be realized about rock is so big and so simple that we are inclined to overlook it – namely, that it is our chief food. If living things were to exist, they could only do so by eating the rocks – there was nothing else for them to eat. And that is just what rocks were and are – food which has been and still is eaten daily. I do not speak figuratively. True, they are not chewed off and eaten like bread or grass, nor when they have become what we might call stone-flour are they taken neat; but their mineral properties, their juices as it were, have always been the basic food of everything, and to this day if we go into a chemist's shop to buy a tonic we will find on the label a list of rock properties to encourage further purchase.

How are the rocks prepared as food? Where are the mills for grinding the stone into 'flour'? Again we must answer that this was the work of the raindrop – and still is.

1. Not included in this volume.

We see this work as having been accomplished in two great movements: one belonging to the past, the other still going on. The first was when the hydrosphere which was held aloft in the atmosphere fell down upon the burning rocks. This meant tremendous chemical combustion, erosion, pulverization, so that the first sea was a strange sea indeed, 'whose waters, boiling hot and thickened with mud and slime of all sorts, formed a kind of mineral pea-soup shrouded in a chaos of steaming vapours'.[2] This mineral pea-soup, this sediment, in due course sank to the bottom, solidified, and became sedimentary rocks. As the aeons rolled by, continents, under the influence of vulcanicity, rose from the sea and the sedimentary rocks became part of dry land, leaving traces to this day upon some of the highest mountains. Yet – and this is what I feel bound to emphasize – this sediment did not necessarily always and everywhere come up hard; it may have come up soft and thus formed the basis for the first soils of the world.

That is what I would call the first movement of the raindrop for turning rocks into soil. The second movement began when the land-blocks and the sea-blocks had sorted themselves out and the earth consisted of dry land as yet uninhabited by organic life. There was no sand – which I have called stone-flour – save what may have been supplied by the sediment caused by the first great invasion of water from the skies. The second movement saw the oceans coming down as rain and snow upon the mountains, which by means of frost-chiselling, and ice-prising, and glacier-work, and direct acid-action of the raindrop, and the sheer power of torrents, brought about the denudation and degradation of the rocks. That was the second great movement. The cold mountains were exposed to the assault of water. It was – and is, for it still goes on – a battle in which the mountains are passive and cannot throw back the assailant as fire did: the victory must go to water, as flake by flake and stone by stone and rock by rock and avalanche by avalanche the mountains are brought low and turned into soil. Thus today every meadow and prairie, every valley and plain, every farm

2. J. H. Fabre, *This Earth of Ours.*

and homestead, and every city throughout the world, rests upon the foundation of ruined rocks.

We realize, of course, that rocks, however much they are ground into sand, are not in themselves soil – they are not loam, not vegetable mould. But they made vegetable mould possible, they served as the foundation for it. At one time the moist, chipped rocks formed the basis for the first plants. Do not ask me how the first plant arrived, or the first animal. Do not ask me how the first organic form entered the lithosphere or the hydrosphere or the atmosphere – for I have no idea. I assume that inorganic life is really organic and that the latter grew out of it, so that there is no real gulf between the dance of the magnets in the molecules and the dancers at Sadler's Wells. I would say with Keyserling:

The music of the universe slumbers in the irrational laws of crystallization; all artistic ideals are symbolically preconceived in the germ of the plasma. From the first breath of desire which trembled through shapeless chaos, an unbroken chain of developments leads to the *Iliad* and the Parthenon.[3]

Anyway they did arrive (and the miracle of their *arrival* should silence the bewildered gentlemen who seem to think that the means of *survival* is of religious import), and having done so they gave themselves up to the further creation of soil. The great work was begun by which the ashes of dead vegetation, and of dead animals, become the life-giving basis for future growth.

2. A Piece of Chalk

... At last the river gets to the sea. Think of the Mississippi. Every minute every day it empties into the Gulf of Mexico over 40 acres of water 20 feet deep. Think of the Amazon. Again, as the second hand of our watches makes but one revolution, that

3. Count Herman Keyserling, *The Travel Diary of a Philosopher*, Vol. II, p. 251.

river unloads a burden of over 100 acres of water from its mouth. It we add to this the water flowing into the sea from the Nile and the Yellow River, from the Congo and the Volga, from the Indus and the Danube, from the Rhine and the Vistula, from the Tagus and the Thames, and from thousands of others, we might momentarily wonder why the oceans are not rising. Then we remember how, while these horizontal rivers feed the seas, at the same time an equal quantity of water is rising vertically into the atmosphere. A balance is maintained: the wheel revolves: the cycle runs full circle.

This does not mean that the oceans do not sometimes vastly invade the land or the land rise from the sea. Hitherto we have been thinking of how the waters shape the landscape and wash away the earth and to a certain extent create further ground with silt. But our thoughts were too parochial to embrace the whole truth, just as they were too functional to include much of the truth of beauty. If we contemplate the scene without personal reference, but in terms of eternity, a deeper vista of what is given and what is taken away by the movement of the waters will be opened before us.

I have already had occasion to quote from the distinguished scientist Dr Kingdon Ward. He is inclined to write in almost a succession of inspired notes, leaving off just when you want him to go on; but he writes underivatively, out of himself, as one having authority and not as the scribes and the pharisees. Speaking of the incalculable changes which have been wrought by rivers in the course of ages, he says:

Through the long ages during which rivers have flowed over the land surface, not only has a great deal of crust suffered demolition and the debris been removed, but it has been reconstructed elsewhere. This has happened again and again, often using the same materials several times over. Land has appeared, has been destroyed, returned to stock for alteration and repairs, to reappear in brand new form elsewhere.[4]

He is referring to the action of vulcanicity, I think.

4. *About this Earth.*

The rivers, as we have seen, carry an enormous amount of sediment into the sea. There it settles down into a bed, compacted under the pressure of the ocean. But it does not always stay there – it may come up again. We know that it has quite often come up again, for the sedimentary rocks bear witness to their marine period. 'The sea has gone back many times to the sea,' said Leonardo da Vinci long before geologists could support him, 'and no part of the earth is so high but that the sea has been at its foundations, and no depth of the ocean is so low but that the loftiest mountains have their bases there.' This is due to the forces of vulcanicity. There are periodic victories of fire over water. The unquenchable heat at the core of earth has again ripped up the crust, altering the configuration of the globe, so that soil which had been carried to the bottom of the ocean becomes dry land again. 'Running water and volcanic action,' says Lyell, 'are two antagonistic forces: one labouring continually to reduce the level of the land to the sea, the other to restore and maintain the inequalities of the crust, on which the very existence of islands and continents depends.' That is to say there is land and there is sea; earth with great mountains, seas with great depths, seemingly fixed – yet not fixed. The tempest rages, the wind beats upon rain-eroded rock; for countless ages the rivers ravage their way to the ocean. A portion of the land goes to the bottom of the water. Then there comes another volcanic action, low places are lifted up even from the bottom of the sea, and new lands formed. Such is the geological story. There is evidence that this has already happened at least four times. Will it occur again in the future? I do not know, but I suppose so.

We are speaking of a tremendous export trade, in the exact meaning that goods do, in vast quantities, make their exit from ports, and in due course there is a return in kind. Sediment is not the only export. There is a huge carriage of minerals in solution. Minerals, extracted from the rocks, are dissolved in water and carried into the ocean. Having reached it they do not remain in solution. One of them conspicuously does thus remain – salt – but others are used in the building up of animal frameworks. This is especially true of calcium carbonate. There is a creature dwelling at the bottom of the oceans called the globigerina. It is microscopically minute, a mere particle of living jelly, devoid of mouth, nerve, or muscles, yet able to eat and grow and multiply, and capable of 'separating from the ocean the small portion of carbonate of lime which is dissolved in sea-water, and of building up that substance into a skeleton for itself'.[5] Billions of billions of these globigerinae throughout all the geological ages have thus built up their bodies and laid down their skeletons. The resultant composition is called chalk.

Such a fact is surprising. Yet a microscopical examination of a piece of chalk will show thousands of perfect shells in one cubic inch. The stratum is a burial-ground. 'The evidence furnished by the hewing, facing, and superposition of the stones of the Pyramids,' wrote T. H. Huxley, 'that these structures were built by men, has no greater weight than the evidence that the chalk was built by globigerinae; and the belief that those ancient Pyramid-builders were terrestrial and air-breathing creatures like ourselves is not better based than the conviction that the chalkmakers lived in the sea.' We look round upon the land of England, and find much of it resting on this foundation, an area covering 3,790 square miles. We can follow it a distance of 280 miles from Lulworth in Dorset to Flamborough in Yorkshire; we find it entering into the foundation of the south-eastern counties, except in the Weald of Kent and Sussex, attaining in some places a thickness of 1,000 feet; it forms our Downs, our Chiltern Hills, our Salisbury Plain, our Beachy Head, our Cliffs of Dover. Beyond these shores it stretches over much of France and the Pyrenees; it runs through Denmark and Central Europe and extends southward to North Africa; it is found in the Crimea and in Syria, and may be traced as far as the shores of the Sea of Aral in Central Asia. It is said that if all the points at which true chalk occurs were circumscribed they would form an area as great as that of Europe.

I have spoken in terms of an export trade. The globigerinae,

5. T. H. Huxley, *On a Piece of Chalk*.

so humble in their individuality and so powerful in their multiplicity, have gone far for their goods. Together with some other
shell-makers, they have looked far afield for their frames, their
mantles, their bones. They have searched the mountains, they
have scoured the plains, and, using the rain and harnessing the
rivers in their service, they have turned their spoils into living
texture of bone and shell. At death they have laid down their
skeletons, layer upon layer, which in time, by virtue of volcanic
action, have been imported back to the land to form the lime-
stones of the world. Thus when we gaze upon the hills and cliffs
of chalk we behold the furthest wildest work of water. We have
seen it carve out caverns in the limestone hills, and now we see
that by feeding these creatures it is the creator of the very
mountains which it carves and hollows – nay, that it is the
bricklayer of the walls of Albion.

CHAPTER IV

SECTIONS FROM
BELIEFS AND CEREMONIES

1. Rain-making and the Gods

If no rain falls and there is a drought the plants and animals perish. They do so without protest and without understanding. Not so with Man. He thinks about it. He cannot bring down rain where and when he wants it (though he is beginning to do even this, as we shall see), and he cannot stop it from coming if there is too much of it; but today he knows what it is, and how to reserve it, and so on. In early days, at the dawn of men's consciousness, it was very different.

It is difficult for us to realize the extent of that difference. The primitives knew so little about Nature that they did not even know that spring came round every year. It came and it faded away, and there was always the hope that it might return, but no certainty. They could see it coming and going year after year, and yet not conclude that it would always do so. It is hard for us now to realize that they could not put that amount of two and two together. The fact is, winter is a long and dreary affair, and it began to seem endless each time; summer was forgotten, and they never thought in reference to a natural law, but only to gods whom they regarded as very capricious. Thus when warmth and fertility did come round again they experienced a tremendous sense of relief and gratitude which they expressed in the most elaborate rituals and wild orgies and cruel sacrifices.

There was nothing more important for the promotion of fertility than rain. Again we must remember that they did not think of rain in any kind of way as a meteorological phenomenon obeying certain laws, but as something given or withheld by the fickle gods – in this case the specific rain-god. By practising certain ceremonies and sacrifices they might be able to prevail upon the god to be good to them, or they could obtain rain by means of sympathetic magic – the idea that your own actions call out a similar response in things and persons around you, and that thus by imitating rain you would get rain. The shrewder members of the tribe would come forward as specialists in this kind. They would claim to have the ear of the deity and to know exactly the most effective forms of magic required. The chief point of Frazer's portion of *The Golden Bough* called 'The Magic Art and the Evolution of Kings' is that certain men did come forward in this way, and when they managed to persuade the people that they possessed the power to bring rain and increase fertility and so on they were able to assume control and be hailed as kings. But to be a king under such conditions was no sinecure; for if he was too often unsuccessful in the fulfilment of his promises the people would turn upon him and kill him. We may wonder how he could ever have been successful save occasionally by chance. I think we are bound to assume that the man willing to claim these powers – the magician – must have possessed some objective observation and have noted that under such and such conditions – say, of mountain and wind – at such and such times rain was inclined to fall; and if he hastened to promise a downpour just before it came his prestige would be high, and might tide him over the periods when he failed. There is also little doubt that, with good priestly organization around him, he could for a long time frighten the people by elaborate threats and insinuations as to the displeasure of the god on account of their sins and inadequate performance of ceremonies and sacrifices.

The persistence, the number, and the detail – sartorial and otherwise – of ceremonies carried out in order to procure rain or avoid too much of it make a study in themselves. Some were extremely grim, as when they dug up corpses, buried children alive, or tortured animals; some curious, as when in order to stop rain certain members of a tribe had their teeth pulled out, or, to promote it, naked women would be harnessed to ploughs and made to imitate the operation of ploughing while invoking the spirit of rain to have mercy on them. There were also more light-hearted ways of going about the business. If rain were needed it was sometimes thought to be sufficient if women seized any passing stranger and threw him into the river, while very good results were supposed to follow if a number of half-naked women took a half-witted man to a river and sprinkled him with water. Throwing dirt or filth at the houses of neighbours was also considered by many to do good, since this would call down upon them curses and vituperation; for there was a belief that it was lucky to be cursed, so much so in fact that fishermen before going out to fish would play a rough practical joke on a comrade in order to be execrated by him, acting on the assumption that each curse would bring at least three fish into their nets, just as no huntsman felt satisfied that he would have a good day unless someone had expressed the wish that he might break his neck. The best vituperations were obtained by drenching the lame, the halt, and the blind, or by insulting shrews, who could always be relied upon to come out with a fluent stream of foul language. In times of drought the good fortune expected to follow from the curses should be in the form of rain.

It is not surprising that they sometimes lost patience with the gods who controlled the weather, and were ready to do them violence. When the drought had been long and their tempers had become short, 'at such times', says Frazer, 'they will drop the usual hocus-pocus of imitative magic altogether, and being far too angry to waste their time on prayer they seek by threats and curses or even downright physical force to extort the waters of heaven from the supernatural being who has, so to say, cut them off at the main.' In their anger on such occasions they have been known to cast down the image of their god and, throwing it into a smelly ricefield, cry, 'There you may stay yourself for a while, to see how *you* will feel after a few days' scorching in this broiling sun that is burning the life from our cracking fields.' The Chinese possessed no more self-control in this matter than others, and once the Governor of Canton in a time of drought, having repeatedly ascended to the temple of the god of rain dressed in his burdensome robes in great heat, burst out, 'The God supposes that I am lying when I beseech his aid; for how can he know, seated in his cool niche in the temple, that the ground is parched and the sky hot?' Whereupon he ordered his attendants to put a rope round his neck and haul his godship out of doors, that he might see and feel the state of things for himself.[1]

Such practices were not confined to the East. The Christian idols came in for rough handling also from time to time in the same way. It used to be the custom to dip the images of saints in water to get rain. In some villages of Navarre prayers for rain were offered to St Peter whose image was carried in procession to the river. At the edge of the water an invitation would be addressed to him three times to send them a good shower, but if he remained obstinate and refused to help he was plunged into the water amidst the curses of the people and the expostulations of the clergy, 'who pleaded with as much truth as piety that a simple caution or admonition addressed to the image would produce an equally good effect'.[2] The most striking case of this kind was in Sicily in the year 1893 when, a drought having lasted six months, everything was withering, food becoming scarce, and the alarm of the people great. Prayers, processions, and vigils had proved in vain, and even the carrying of crucifixes through the wards of the town and the scourging of each other with iron whips did no good at all. Masses, vespers, concerts, illuminations, and fireworks could not move St Francis. At last all patience was exhausted. Some saints were banished. At Palermo St Joseph was put outside to observe the state of the weather for himself. Other saints were turned with their faces to the wall, like naughty children. Others were ducked or insulted, while St Angelo was put in irons and threatened with drowning or hanging, 'Rain or the rope!' roared the mob, shaking their fists in his face.

When we review the practices carried out in the name of religion or magic we are sometimes moved to think that what was often really at work was the age-old unconscious sadism

1. S. Wells Williams, *The Middle Kingdom.*
2. *The Magic Art.*

that informs so much of human psychology. Thus, the inhabitants of Buru are threatened with destruction by a swarm of crocodiles. What is wrong? The prince of the crocodiles has a passion for a certain girl. So her father is compelled to dress her in bridal array and deliver her into the clutches of her crocodile lover. Did they really believe this or just enjoy it? It is impossible to say – for certainly no belief is too fantastic to be believed by people in any age. It was the custom of the Egyptians to drown a girl on the rise of the Nile every year, until an Arab general abolished it. In the same way the Chinese used to marry a young girl to the Yellow River once a year by drowning her in it. The witches chose the fairest maiden and superintended the affair. A local mandarin at last came forward and forbade it. This deeply pained the old women. It was the custom. It had always been done. What would the people say at this scandalous breaking of a time-honoured tradition? Determined not to be baulked of their fun, they ignored the edict and made the usual preparations for the murder of the girl. Happily, the magistrate called out the soldiers and had the women bound and thrown into the river to drown instead, assuring them that the god of the river would no doubt be pleased to choose one of *them* for bride.

Indeed, China often provides us with a lighter and more urbane note in these matters. It is agreeable the way they made no bones about taking heaven by storm if the gods were not doing well. Sometimes they threatened to beat the god if he refused to send rain, or to depose him from the rank of deity, while if the needed rain fell they were ready to promote him to a higher rank by imperial decree. A long drought desolated some provinces of northern China in the Manchu Dynasty. Processions failed to move the rain-dragon. The Emperor's patience became exhausted, and the deity was condemned to perpetual exile. The decree was in execution when the judges of the High Court of Peking became so sorry for the poor dragon that they implored the Emperor to pardon him. The Emperor was touched and revoked his doom, and a messenger was sent off at full gallop to bear the tidings to the executors of the imperial justice – and the dragon was reinstated on condition that he would do better in future. In the same manner, in April 1888, when at Canton there was an incessant downpour of rain, the mandarins requested the god to stop it. Failing which, they locked him up for five days. The rain ceasing, he was restored to liberty.

Right into the nineteenth century it was considered the duty of Chinese officers to secure genial seasons by their good administration, and if bad harvests or epidemics ensued the fault was found in them. They were sometimes obliged to resort to some risky expedients. In 1835 the Prefect of Canton, on the occasion of a distressing drought which had lasted for eight months, appealed to a rain-maker. 'I respectfully request him to ascend the altar of the dragon and sincerely and reverently pray. And after the rain has fallen, I will liberally reward him.' A man came forward, vainly repeating incantations from morning to night, exposed barefooted in the hot sun, the butt of the jeering crowd. The Prefect himself was lampooned. Finally all slaughter of animals was forbidden, a fast proclaimed, the South Gate closed, all prisoners released, and 20,000 people prostrated themselves before the shrine of the Goddess of Mercy. At last rain came. They made thank-offerings, the South Gate was opened, and they burnt off the tail of a live sow. The officers and literati, though acknowledging the folly of these observances and ridiculing the worship of senseless blocks, nevertheless still joined in, and during a severe drought at Peking in 1867 supported the belief that the sacrifice by the Emperor of a white tiger to the dragon would be sure to liberate rain.

The Emperor, in his character of Vice-Regent of Heaven, felt himself to be still more responsible for natural calamities. During the drought of 1817 the Emperor Kiahing said, 'The remissness and sloth of the officers of the Government constitute an evil which has long been accumulating. It is not the evil of a day; for several years I have given the most pressing admonitions on the subject, and have punished many cases which have been discovered, so that recently there appears a little improvement, and for several seasons the weather has been favourable.' The most remarkable imperial prayer, or memorial, presented to Heaven by the Emperor is that drawn up by Taukwang on 24 July 1832, on the occasion of a severe drought at the capital. He had tried many things in vain, and felt bound as high priest of the Empire to show the people that he was mindful of their sufferings and would relieve them if possible by presenting a memorial to Heaven. It begins:

Kneeling, a memorial is hereby presented, to cause affairs to be heard.

Oh, alas! imperial Heaven, were not the world afflicted by extraordinary chances, I would not dare to present extraordinary services. But this year the drought is most unusual. Summer is past and no rain has fallen. Not only do agriculture, and human beings feel the dire calamity, but also beasts and insects, herbs and trees, almost cease to live. I, the minister of Heaven, am placed over mankind, and am responsible for keeping the world in order and tranquillizing the people. Although it is now impossible for me to sleep or eat with composure, although I am scorched with grief and tremble with anxiety, still, after all, no genial and copious showers have been attained.

Some days ago I fasted and offered rich sacrifices on the altars of the gods of the land and the grain, and had to be thankful for gathering clouds and slight showers; but not enough to cause gladness. Looking up, I consider that Heaven's heart is benevolence and love. The sole cause is the daily and deeper atrocity of my sins; but little sincerity and little devotion. Hence I have been unable to move Heaven's heart, and bring down abundant blessings.

Having searched the records, I find that in the twenty-fourth year of Kienlung my exalted Ancestor, the Emperor Pure, reverently performed a 'great snow service'. I feel impelled by ten thousand considerations to look up and imitate his usage, and with trembling anxiety rashly assail Heaven, examine myself, and consider my errors; looking up and hoping that I may obtain pardon.

He then examines various possibilities wherein he may have fallen short; whether by imperfect and disrespectful sacrificial ceremonies, or remissness in attending affairs of State, or the utterance of irreverent words, or the performance of imperfect justice, or in extravagance, or in faulty appointment of officers, and so on. Finally he winds up like this:

Prostrate, I beg imperial Heaven to pardon my ignorance and stupidity, and to grant me self-renovation; for myriads of innocent people are involved by me, the One man. My sins are so numerous it is difficult to escape from them. Summer is past and autumn arrived; to wait longer will really be impossible. Knocking head I pray imperial Heaven to hasten and confer gracious deliverance – a speedy and divinely beneficial rain to save the people's lives and in some degree redeem my iniquities. Oh, alas! imperial Heaven, observe these things. Oh, alas! imperial Heaven, be gracious to them. I am inexpressively grieved, alarmed, and frightened. Reverently this memorial is presented.[3]

It is an interesting document. It was a bit of a gamble. It would hardly have done for His Imperial Majesty, addressing himself as crowned head to crowned head, to have been ignored when the memorial was received in Heaven. Happily, he was spared the ridicule of his people, for heavy showers followed the same evening.

2. Detection by Water

. . . Apart from their preoccupation with rain-gods, we find that waters, or certain waters, were regarded as sacred. They felt them to be instinct with divine life and energy. There were all sorts of legends connected with sacred waters. They were associated severally with various deities and supernatural powers from which the people felt that they drew vital energy. 'Among the ancients blood was conceived as the principle and vehicle of life, so the account often given of sacred waters is that the blood of the deity flows in them.'[4] This was often apparently supported by the ruddy colour of the soil.

They drew not only energy but knowledge. They believed

3. S. Wells Williams, *The Middle Kingdom*, Vol. II, p. 468.
4. Robertson Smith, *The Religion of the Semites*.

that when they drank certain waters they could learn certain things: thus, not so long ago, at Marsala in Sicily, at the Grotto of the Sibyl, at the ancient site of the Temple of Apollo, on St John's Eve, 23 June, women and girls used to visit the grotto with the belief that by drinking of the 'prophetic water' they would learn whether their husbands had been faithful to them in that year, or whether they would be married in the year to come. There are many instances of water being associated with oracles. Robertson Smith suggests how this came about. At first, at Mecca and at the Stygian waters in the Syrian Desert, for instance, worshippers threw gifts into the holy source. Since the worshipper liked to receive some visible indication as to whether or not his prayer was accepted, he was ready to consider it as being accepted if his gift sank, and unacceptable if it was cast forth by the eddies. Thus the holy well, by declaring favourable or not the disposition of the divine power, became a place of oracle or divination.

This led to a further stage. Certain waters or wells became places of detection and ordeal, where an accused person could be tested by being presented at the sanctuary, on the assumption that no impious person could come before the divine god with impunity. This idea was brought forward into the Middle Ages in the form of trial of witches by water, when a man claiming injury by enchantment would be allowed to have all the suspected witches brought to the sea or a deep pool, and then with heavy stones tied to their backs be cast into the water. The witch who did not sink was regarded as the guilty person, since it was certain that the sacred element would reject the criminal. The guilty witch would then be burnt, while the others enjoyed exoneration from all blame – though unhappily drowned.

If impure persons could not safely approach holy waters, still less could they drink of them. For such a person to take them into his system would mean disease and death, and Robertson Smith tells how 'at the Ashbamacean lake and springs near Tyana the water was sweet and kindly to those that swore truly, but the perjured man was at once smitten in his eyes, feet, and hands, seized with dropsy and wasting. In like manner he who swore falsely by the Stygian waters in the Syrian Desert died of dropsy within the year.'

That was in reference to the Hebrews. In the cosmogony of the Egyptians and Babylonians water was regarded as the original home of the gods, the father of all gods. He was EA, source of all wisdom and learning. And there the divination by water seems to have been equally elaborate. There were many sorts of 'divining cups', and many sorts of divinations. An inquirer would ask the magician a question, the magician would look into the cup or bowl, and see, or profess to see, the scene in question enacted on the surface of the water. And in terms of detection the following example is particularly interesting as belonging practically to our own times in date. At Khartoum a native on going to Cairo deposited some money with a friend for the use of his wife. On his return the friend denied that he had given him any money and swore that he was mistaking him for someone else. The man went to the Kâdî, who thereupon decided to test them both by means of the divining bowl. Writing upon a sheet of paper words from the Koran he then dissolved it in water in an earthenware bowl, and commanded both men to drink, each to take half of the water, warning them that the judgement of Allah would fall upon the liar after he had drunk his portion. Each man took his draught. The man who had received the money died forthwith. But this did not satisfy his friends, who cried 'Poison!' and accused the lender of the money of murder, and had him hauled off to prison, and even caused a guard to be set over the house of the Kâdî, who was suspected of complicity.

In due course the man was formally charged with poisoning his friend, and the British authorities were inclined to believe him guilty. But so many representations were made by the Muhammadan elders, who declared that the death was due to the effect of the words of the Koran dissolved in water, that the case was referred to Lord Cromer in Cairo, and a judicial decision demanded. Cromer, a man of comprehensive mind and broad sympathies, laid the matter before the *mullahs* of the Ay-Hyhar University in Cairo, who after exhaustive inquiry gave it as their unanimous verdict that the man who fell back dead was a liar and a thief, and that it was Allah Himself who had killed him by virtue of the Koranic words dissolved in the bowl. Lord Cromer bowed to their decision and the man was released.[5]

3. Water as a Means of Purification

Early in history the peculiarity of water was discovered by man, an attribute not religious nor prophetic nor sacred nor magical, but simply physical – namely, that it had the power to *remove* substances which had become attached to other substances. If your hands had become ingrained with earth, or your clothes covered with it or other things, then by using water – especially if mixed with some form of fat – you could remove it and get your hands and your clothes back as they were before. This is called washing.

It is a common sight to see someone washing his hands. This capacity to wash ourselves, to keep ourselves *clean*, is a tremendous thing, and entirely due to water. Throughout the ages it has had a marked effect on human personality. Before the recent advent of the classless society the difference between classes was instantly recognizable by the quality of the skin, which originated in the greater facility for washing amongst the rich. Indeed, a history of washing would not make a dull book. I can only concern myself here with one point – namely, that the purifying of the body easily became associated with purifying the soul. There came a time when people actually said that cleanliness was next to godliness. Personally, I have never been very fond of that remark. I have often been struck by the spick-and-spanness of, say, an English agricultural labourer's cottage as against the lack of this in an Irish cottage; and with the admirable cleanliness of the one I have found an excessive materialism hard to endure, while with the other I have encountered religious feeling and some spirituality. And one

5. See 'Divination by Water, in *Amulets and Superstitions*, by Sir E. A. Wallis Budge.

recalls how a fourth-century pilgrim boasted that she had not washed her face for eighteen years for fear of removing the baptismal chrism. Still, I do not wish to swing to the other side: I would not believe much in the holiness of any 'holy man' who was dirty in his appearance, nor am I on the side of that pilgrim; I merely observe in passing that this particular catch-phrase, like so many others, leaves room for improvement. My immediate point is that the physical property of water in purifying the body suggested that it also purified the soul.

In earlier civilizations a surprising number of things were believed to have the effect of making a person so impure that it was dangerous to come near him. He was considered to be stained both physically and spiritually, and in need of purification, if he had been in contact with a dead body, with childbirth and menstruous women, with murder and any form of bloodshed, with persons of inferior caste, with dead animal refuse, leprosy, madness, and disease generally. In order to cleanse a person from impurity thus acquired, a number of purification rites had to be performed. Such beliefs have played a large part in primitive rituals, and the various elaborate forms of purification in Babylon, Egypt, and China, amongst the Greeks, the Hebrews, the Hindus, the Iranians, the Jains, the Japanese, the Christians, the Muslims, the Romans, and the Teutons become wearisome to contemplate.[6]

The ablutionary liquid used at these ceremonies was not always water – not at first. Blood, especially the blood of pigs, was a favourite property in cleansing from pollution. Far more widespread than blood, however, was the use of urine. In many lands throughout many ages urine was regarded as a cleanser from impurities, as a warder-off of the 'evil eye', as a bath-water, as a skin-purifier, and as a drink. The veneration extended towards it had the same significance as that extended towards salt, that surprising rock whose property of dissolution in water and capacity to cleanse the putrefactions and corruptions of the living world made it a special object of magic and holiness, and our language is full of superstitious references to it, often in

6. See 'Purification' in *Encyclopaedia of Religion and Ethics.*

terms of erotic – that is, *vital* – association, so that even an adjective like 'salacious' (to mention but one out of so many words in this connection) reminds us of its origin. In the course of his formidable essay on 'The Symbolic Significance of Salt'[7] Dr Ernest Jones observes that 'All the evidence, from comparative religion, from history, anthropology and folklore, converges to the conclusion, not only that the Chrisitan and other rites of baptism symbolize the bestowment of a vital fluid (semen or urine) on the initiate, but that the holy water there used is a lineal descendant of urine, the use of which it gradually displaced'; and he follows with exhaustive documentation to establish the point.

No doubt the erotic, the vital meaning in the ceremony of baptism is most fundamental; but we must not overlook the quite simple and profound appeal which water made and makes in its own right as our life-need. 'Take this,' run the words in the ritual of a Mexican pre-Christian baptismal ceremony: 'Take this: by this thou hast to live on the earth, to grow and to flourish; through this we get all things that support existence on earth; receive it.' Nor must we forget the appeal which pure water makes as a mysterious and holy element with its extraordinary ablutional powers:

> What if this cursed hand
> Were thicker than itself with brother's blood,
> Is there not rain enough in the sweet heavens
> To wash it white as snow?

It is easy to come to think of water as a means not only of ablution, but of absolution; and when we wish to aknowledge a new member into the brotherhood of man by giving him a *name*, when we would admit him into the membership of the Church, what more natural than that we should use water as the sign of grace, the token of his initiation into a purer life and the remission of his sins?

The earlier Christians felt that in order to receive full benefit complete immersion was necessary, and the practice continued

7. In *Essays in Applied Psychoanalysis.*

for many years, as the large baptisteries in our churches, containing ponds big enough for the purpose, bear witness. As time went on, and infant baptism was introduced, sprinkling with water began to be considered sufficient – though the introduction of infant baptism at all led to raging controversies between Luther and Zwingli, Calvin and Hooker and others as to whether it was a means of grace, a seal of regeneration, or just a rite of adherence. And since at all times there are zealous men who take things more seriously and adhere to doctrines more consistently than their fellows, so here a body of men calling themselves Baptists came forward, holding that the rite should be administered to believers only, and that total immersion was imperative. It was, and is, a formidable body – a visitor will be struck by the distinguished character of the pastors and their happy relation with their flock – containing further sects within itself, including Daniel Parker's Two-Seed-in-the-Spirit predestinarian Baptists, still going strong, who appear to believe that a certain number of people are saved from the start, however bad they may be, and a certain number damned from the start, however good they may be.

When we pass in review the number of ceremonies in many lands carried out in the name of purification; of the immense influence which water held for the Nile-conscious Egyptians; of the Greek philosophy of purity with the attendant cathartic rites; of the complicated Hindu practices; of the desperate Iranian protection against the impurity of contact with the dead; of how, even in the Confucian *Analects*, there is a reference to the custom of washing the hands and clothes at some stream in the third month in order to put away evil influences – we realize again the great part which outward shows and symbols have played and perhaps must always play in encouraging simple souls towards some sort of reverence and discipline. Yet it is always pleasant to encounter those who are strong enough to look within for virtue and faith. We recall with relief how the Taoists, before Taoism sank into a hotch-potch of magic and mummery and demonology, displayed a sublime contempt for the popular rites going on around them, and strongly contrasted the 'fasting (or purified) heart' with ceremonial purifications which may be merely fictitious and external, unnecessary and meaningless. 'The true sage does nothing,' they declared, and were careful to add, 'therefore there is nothing that he does not do.' And there is surely something delightful in the attitude taken up by the enlightened Buddhists in the midst of their complicated rituals – as exemplified in the following dialogue:

THE BUDDHA: What of the river, Brahmin, what can it do?
THE BRAHMIN: Many consider it as a means of deliverance and of merit; many people let it bear away their evil deeds.
THE BUDDHA:
    What boots the Bahuka, or the Gaya?
    For ever and a day his foot may plunge
    Therein, yet are his smutty deeds not cleansed.
    They will not purge the man of passions vile.
    To him that's pure, ever 'tis Phalgu-time,
    To him that's pure, ever 'tis Sabbath-day,
    To him that's pure and in his actions clean,
    Ever his practices effectual prove.
    Here, Brahmin, is't that thou shouldst bathing go:
    Become a haven sure for all that breathes;
    Speak thou no lies, harm thou no living thing,
    Steal naught, have faith, in nothing be thou mean.
    So living, what are river-rites to thee?[8]

Such words fall pleasantly upon a modern ear. Yet I would not wish to deny the truth in the conception of water as holy and with holy powers, nor give less than full value to Keats's words: 'The moving waters at their priest-like task Of pure ablution round earth's human shores.' Symbolism, of course, has its place in poetry and its place in religion, and maybe no earthly vesture serves better than water. I would add that it also serves well as a symbol in more secular terms. Such is provided by means of comedy (no less profound for that) by Charles Dickens in the figure of Jaggers in *Great Expectations*. Jaggers, the stern lawyer, saw much of the unseemly side of life; and the

8. *Majj-bima-Nikaya*, I. 39.

purificatory rites which he considered necessary are interesting. 'He washed his clients off,' says Dickens,

as if it were a surgeon or a dentist. He had a closet in his room, fitted up for the purpose, which smelt of the scented soap like a perfumer's shop. It had an unusually large jack-towel on a roller inside the door, and he would wash his hands, and wipe them and dry them all over this towel, whenever he came from a police-court or dismissed a client from his room. When I and my friends repaired to him at six o'clock next day, he seemed to have been engaged upon a case of darker complexion than usual, for, we found him with his head butted into this closet, not only washing his hands, but laving his face and gargling his throat. And even when he had done all that, and had gone all round the jack-towel, he took out his pen-knife and scraped the case out of his nails before he put his coat on.

CHAPTER V

SECTIONS FROM *THE SERVANT OF MAN*

1. The Paradox and Poetry of Drainage

The story of man and Nature is by no means an inglorious one. Not at first, anyway. He came, he saw, he imagined, he dared, and he wrought. He applied his imagination. Applied imagination is called engineering. When we contemplate some of the mighty works of engineers and architects, we are tempted to call them applied poetry.

From thousands of years B.C. the capacity of man to make the forces of Nature serve him beneficently has been exemplified – especially in relation to water. Though men have made great efforts to persuade the sky to yield rain, they have not been content to wait upon it. From the earliest times they have spread water over the land where there has been no rainfall. They could do this wherever there are rivers.

We have thought of the ocean as being 'on loan' when it comes down on the earth. We see it flowing back in the form of rivers. It is not much use to us that way – not as rainfall. But it can be made to serve as rainfall. A river can be made to spread out all over the place and water the land in rainless regions. It can be made to irrigate the land.

We at once think of the Nile. And indeed it is the classic example of a beneficent river. All rivers could hardly be thus described. The Congo, for instance, is so wide at places that you cannot see across it. From a functional point of view this is a frightful waste – the ocean vastly flowing back to the ocean without doing its job for us. It can, of course, be used for irrigation, but how much more satisfactory it would be in the form of hundreds of small rivers instead of one huge highway. The Nile is a different story. It might even be called the teacher of irrigation, for by flooding the land at punctual periods it showed how rivers could be used as rainfall. The Egyptians were not slow to realize how they could channel and control these periodic floods. Having entered Egypt, the Nile flows through the land for 700 miles without receiving a single tributary. It does not receive tribute, but it gives tribute. Waters do not flow into it; waters flow from it every year, making the Valley of the Nile famous for its fertility during hundreds, even thousands of years. The Egyptians belong to an almost cloudless and rainless land. But they need not pray for rain. They need not look up into the empty heavens hoping to behold the coming of a cloud. For they do receive rain: they do have their cloud – flowing beside them in the form of Old Mother Nile.[1]

As it was yesterday, so it is today: 'the Nile is the dominating feature of the north-east quarter of Africa,' Mr H. E. Hurst reminds us, 'and affects all of Egypt, the Sudan, and Uganda, one-third of Ethiopia, parts of Kenya, Tanganyika, Ruanda-Urandi and the Belgian Congo.' Those simple words suggest – do they not? – how much one river can do for life on earth. Mr Hurst finishes his thought with the following sentences which go well with the image which I have striven to hold up regarding the universal revolution of the wheel:

Its annual cycle is produced by the prime source of all our energy, the sun, which evaporates water from the South Atlantic Ocean, and this, by pressure differences of the earth's rotation, is driven 2,200 miles across Africa to the highlands of Ethiopia, where it falls as rain. The water runs off the hillsides scouring away much earth, and runs by little rivulets into hundreds of little streams which find their way into the Sobat, Blue Nile, and Atbara, and so into the Mediterranean, to be again drawn up into the atmosphere and again circulated to who knows what part of the earth.[2]

The Nile offers two exceptional services. In the first place, at

1. In old times they did offer sacrifices to the river and prayed to the god: and Frazer adds that the Emperor also took the precaution of throwing into the river a written command that it should rise – which it always obediently did. (Perhaps he knew it would, and sought to add to his prestige.)
2. *The Nile*.

punctual periods, on account of the melting of the snows hundreds of miles away, it overflows, and can be carefully guided into a wonderful network of irrigational canals and basins. In the second place, by doing this, it also empties a large amount of mud, of silt, over the land, thus offering a yearly supply of fresh soil – in short, doing the positive rather than the negative work of rivers, the alluvial work of taking soil from one place and laying it down at another place rather than carrying it out to sea. This second service cannot be emulated by man: he cannot supply himself with fresh soil in this way from other rivers. He can make them flood over. By causing a given river to flow out into prepared canals and ditches and basins and reservoirs he can make good use of the water going to the sea and make it spread out over his land as if it were rainfall. Stated in the simplest possible terms, that is what irrigation amounts to: and the history of irrigation celebrates the watering-can process of splashing the liquid over the fields from these artificial channels – the extent of the progress being measured by the difference between the primitive *shadûf* (still used)[3] or simple water-wheel at one end and the elaborate hydraulic apparatus and spraying machines at the other; and from water reserved in the carved-out trunks of baobab trees to the enormous reservoirs of modern times.

Though we can state quite simply what irrigation is, and so easily take it for granted, we should not forget that it was once a great *discovery* and came as a marvellous revelation of mercy, akin to the revelation of the first harvest. It altered history. It made history. No longer was it necessary for men to be nomadic; they need not pass from or perish on parching lands – they could settle for ever. Irrigation spelt civilization. Thus Egypt's great high lamp was raised, and its light of thought, of mathematics, of geometry could illuminate the world for centuries to come.

We have also the right to be proud of our efforts in the opposite direction. Every land does not need irrigating, for some get enough water. But all lands – even those which have

3. See p 69 of Raven-Hart's excellent *Canoe Errant on the Nile*.

received irrigation – at times get too much water, and steps have to be taken to prevent a useless mulch of mud or sour water-logged soil. These steps are known as drainage – another great task to which man has applied his knowledge and skill. The principle of drainage is to enable plants to receive water, but not too much. Since rain is so inconsistent and so often comes in short rushes instead of steady delivery, and since irrigated supply cannot in the nature of things be an equable spray, there is always danger of clogged-up soil. What is more, if crops are given too much water they will get too little water. If there is plenty of moisture on the top layer they will not strike their roots deep, and thus may perish in time of drought. But if the soil is equalized and deepened by drainage the roots will go much farther down. Here, as always, ephemerality accompanies shallowness, and endurance depth; and we see again and again how the most glorious vestures reaching up to the loftiest bright-ness have the deepest roots that explore the darkness. This is the paradox and poetry of drainage – and its most eminent phil-osophers are fully aware of it[4] – that its main purpose is to see that plants get more water by being given less.

Thus we have the pleasing spectacle here of seeing man as an agent of the creative forces, helping to bring about increase of life. The crops which decorate this earth with so much beauty would not flourish – nay, would scarcely exist – without man's aid. Failing our attentions, there would be a confusion of rank growth struggling for existence, with no quarter given to delicate grasses and grains. The less robust plants are unable to flourish if there is too strong a solution of salts in the soil – and one of the tasks of drainage is to lessen the percentage. An abundance of free oxygen is indispensable to the life of the plant: the seeds must have it, or they rot; the roots must have it, or they cannot do their work; the soil bacteria must have it, or they cannot supply the plants with nitrogen. It is the task of drainage to guard against stagnation of soil water, which often

4. As a striking illustration of simple experiment see Fig. 2 in *Soil*, by G. V. Jacks.

results in the exhaustion of oxygen. Finally, there are hundreds and thousands of acres which are water-logged to start with – the swamps of the world. It has been one of man's major triumphs to turn swamps into fertile land. There is nothing more depressing than a swampland. It offers nothing to the eye, nothing to the spirit. How wonderful then to transform it into a thriving place. A great task, calling for courage and patience and skill. When we think of what has been done from time to time in this way, both in the New World and the Old, we may appreciate that passage of Goethe's in which Faust discovers that the perfect moment – the moment when he would command all the clocks in the world to stop – was not when he was with Helen or Gretchen, but when he was organizing the draining of the marshland, and establishing a new community where men could lead the good life.

2. Water-divining

... Before considering further services of water there is one thing we should try to clear up. When the water is flowing past us or springing up before our eyes we have merely to help ourselves to it. Even when it is not in front of us we dig wells and come at it that way. But how do we know where to sink wells? There is a vast quantity of ground water, as we have seen, but it is not to be found everywhere. Before we go to the labour and expense of digging a well we want to be sure of there being water at that place. Since it is out of sight underground we cannot discover it in the ordinary way of searching for something. To a remarkable extent geologists can successfully predict the presence of water. When such deduction proves wrong we must divine where it is. This is beyond the capacity of most people. Luckily a certain number of men actually do possess the power to divine the presence of water, and we call them water-diviners. They walk along where it is hoped that water may reside, holding in the fingers of both hands a little instrument (as if it were a wand), generally a hazel twig, which seems to become alive and twists about when there is water to be found underneath.

The existence of such men has been acknowledged for centuries – ever since Moses struck that rock, we might say. They said, 'Water is *there*,' and water was there. That might seem enough to establish their case and prove their authenticity. But in all ages there are strong-minded persons – and a good thing too – called sceptics, who question the authenticity of anything which is not easily explained. Water-divining is not easily explained and some of the explanations advanced by its practitioners (who did not necessarily understand their gift) have been feeble or phoney; so the sceptics insisted that water-divining was a fraud or a fake, that no such thing was possible, that it was not proven, that it didn't really happen (merely passing through the hollow form of actually occurring). The years passed and more water-diviners, or dowsers, went on finding more water, until at last scepticism was dropped and their mystery now qualifies as a science and has been given the more dignified name of radiesthesia.

Before attempting to unravel the meaning of that word let us take an example of a successful diviner before these days of scientific recognition. The case of B. Tompkins will serve, a nineteenth-century man. It was at first with great reluctance that he acknowledged his gift, which he felt as an unwelcome guest dwelling within him. His description of his first experience tallies with the experience of others. 'I was in an orchard,' he writes,

where there was no sign of water or wells or any indication that water existed at one place more than another. After walking a distance of eighty yards or thereabouts I suddenly felt a running or creeping sensation come into my feet, up both legs and through my body, down my arms, causing the rod to begin to rise in my hands. I gripped it still tighter and kept walking, and the feeling became stronger, and I felt I was being led in a zigzag course, at the same time the twig exerting a stronger determination to turn over. So strong did this influence become that I was powerless to keep the twig down, and eventually, after proceeding some distance further, it attained a vertical position and revolved over and over. So great was this sudden and unexpected pressure or influence on me that I fainted and became very ill. I threw the rod away thinking Old Harry was not far off.

It was not till two years after this that Tompkins reluctantly yielded to using his gift in a professional capacity. Meanwhile, he had tested himself, hoping *not* to be right, but it was no good: 'adopt what course I would, my tests always came out right,' and at last he felt compelled to become a professional water-diviner. At the beginning of his *Springs of Water. How to Discover Them by Divining Rod* (from which I have just quoted) he lists 270 patrons – today he would have called them clients – including graces, lords, sirs, and honourables; captains, colonels, generals, and clergymen; estate agents and parish councils; architects and surveyors; land companies and laundries; solicitors, engineers, and manufacturers; dairy concerns, breweries and railway companies; waterworks and colleges. It is an arresting list, giving one a good idea of the variety of interests needing water and being supplied with it from the hitherto untapped sources below. For Tompkins never failed to find water for these patrons who called on his services. Others, equally in need, refused out of prejudice to call on him, their intellectual need for established fact being greater than their physical need for water – which they preferred to go without rather than obtain by employing a diviner.

The physical distress to which Tompkins was subjected when over the site of running water is quite usual with diviners. We hear of the nausea, the headache, the muscular spasms, staggers and cold shivers of which they are often the victim. We learn of a Cornishwoman who works without a rod because on reaching the spot where water runs beneath her feet her face becomes contorted and her whole body suffers convulsions, reminding us of Miss Penrose, who when she was over a silver-mine in Canada felt such a knife-like pain in her feet that she thought she had trodden on a rattlesnake; while some dowsers are so sensitive that they do their divining from an aeroplane. Such people are slightly bewildered when they are told by the sceptics that they are 'pretending' and that it is all 'superstitious nonsense'. However, that view is now considered out of date, and dowsers have received the blessing of science and the support of governments, with the result that more than one officially employed practitioner in France has been obliged to open offices and form a company.

Indeed, the French seem to be leading the way in this matter. There is a list of sixty-seven books and pamphlets, drawn up by La Maison de la Radiesthésie in Paris, all published since 1940, on the subject. The scope has widened considerably. It was always wider than some people think. Already Tompkins could define dowsing as 'the tracing of boundaries, murderers, mines, metals, minerals, and hidden sources of water'. He was very good at finding gold and was put to all manner of tests, when golden sovereigns were hidden in rooms and even in fields, which he always found, to the amazement of the spectators. A pleasing modern example of this is that of a French radiesthesist who found a box of jewels hidden during the Occupation by its owner, who afterwards suffered loss of memory and did not know where to look. The scope is still wider now, and better organized. It is said that the French police regularly call in radiesthesists to help baffled detectives in tracking down mysterious criminals (would this meet with the approval of Monsieur Maigret?). And is it easier, we may ask, to trace a bad man than a good man, and if so, why? Missing bodies are found regardless of moral status, as in the case of Mr A. Johnson of Northallerton, who was missing from a fishing party on the River Ouse in May 1952. After three and a half days' search, the police gave up and a dowser found the body after one and a half hour's search. Further than their assistance to the police is their assistance to doctors in diagnosing diseases and bringing fresh hope to invalids of whom the doctors had despaired. Stranger yet is the fact that the modern diviner can sometimes find the object without going to the spot – without leaving his room! Thus the Rev. H. W. Lea-Wilson, an Englishman, claims to get the same reactions from a map spread out in front of him as from the prospecting place itself, and is so little affected by distance that when he was in India he used a map to find water for a friend 3,000 miles away, in Wales. The Abbé Mermet, the most celebrated of French radiesthesists, could find missing people simply by using a map of the district where they had disappeared and either a photograph or an object connected with them, and he declared: 'Since sick organs, whether they be those of a man or an animal, do not radiate in the same way as healthy ones, I can discover the seat and the nature of sickness even from a photograph of the sufferer.' (I quote that remark since he made it and could do these things, without understanding his *sequitur*, since we cannot suppose that a photograph of a sick person would throw off tell-tale rays.) Finally, we may take the agreeable story of Monsieur Calte, a popular Parisian radiesthesist, who tells how one morning a girl came to him in great distress, for, having fallen in love with a man she had met on the previous day, she found that she had lost his telephone number and could not remember his name or address. Monsieur Calte, merely requesting the girl to concentrate on the thought of her lover, spread out a map of Paris, and by the help of his pendulum was able to point to a hotel where in fact she found the young man that very afternoon.

It will be observed that Monsieur Calte used a pendulum. This is now used more often than the time-honoured rod. It gyrates clockwise when over the position of the object. It has its disadvantages and may go out of fashion, but at present it is the most popular instrument, and Ronald Mathews assures us that it is very widely employed in France by all sorts of people as well as the professional diviners; by housewives to verify the freshness of eggs or the safeness of fungi; by collectors of antiques to pick out the genuine stuff; by bank managers suspecting a forgery; by works managers to check the purity of metal delivered to their factories; and by up-to-date burglars unwilling to crack a safe without first ascertaining the value of its contents.[5]

5. See *Realities:* Radiesthesia; 'An old hoax or a new science?' May 1952, by Ronald Mathews.

Is there a comprehensive explanation for all this? I think not. In a formidable and exhaustive work called *The Physics of the Divining Rod*, by J. C. Maby and T. B. Franklin, a book bearing every sign of experienced knowledge, the physiological aspect is treated, but not the psychological. That is to say, the authors feel themselves competent to account for a good deal of the problem, but not the divining from a distance, the tracing of missing persons, and some of the other things I have cited; these divinations, they think, can only be explained psychologically, including perhaps a telepathic interpretation. They do not feel able to clear the matter up from that side; and though psychic research happens to be one of their specialities, they are so put off by the number of 'cranks and charlatans who flock round dowsing as round Spiritualism' that they have addressed themselves only to what can be explained from a purely physiological basis. In fact, they divide the subject into two categories: one they would call *divination* the psychic faculty; the other they would call *dowsing*, the physical faculty.

There was a time when virtue was thought to reside in the indicator held in the fingers – the rod. The manner in which it insisted upon turning over in the man's fingers made it appear as if it were reacting on its own. This can be proved to be a fallacy by simply giving the indicator the chance to act on its own hook – that is, by mounting it on a mechanical carrier, when it will be found to register no reaction whatever, so that is out. The gift lies in the dowser himself. He happens to be sensitive to rays emanating from the object: that is why dowsing is now called radiesthesia. His reactions result from 'a physiological reactivity of living tissues (neuro-muscular groups especially) more strongly developed in some subjects than in others'. The genuine dowser must possess this sensitivity in the first place, but to become a skilful practitioner who goes about the job competently and is able to make reports in fairly exact terms of quantity and quality, to make 'a tolerably precise and reliable science out of a simple physiological reflex action of the muscles in response to delicate physical stimuli', is, as Maby and Franklin emphasize, no mean achievement.

Those authors hold that 'physiologically the dowsing reactions need no longer be obscure or mysterious', and in the course of their long and learned study they claim 'incontrovertibly that causes of ordinary dowsing reflexes and rod reactions are to be found in certain penetrating, electrically exciting rays: one class – the more important of the two – consisting of short Hertzian waves of geophysical and cosmic origin, and exhibiting polarization and electro-magnetic phenomena'. Yet if they are here saying, in effect, that the dowser is just a sensitive plate, it is still open for others to maintain, as Ronald Mathews asserts, that 'far from being a mere sensitive plate, he is a radar, which sends out a signal and registers a response which it has itself elicited.'

It seems clear to me that the matter is not yet clear. The field is still open to speculators, and I am not prepared to hand myself over wholly to any of them – especially if they won't tell me why, if rays come from this and that object, they do not come from all objects and hopelessly confuse the dowser. Another specialist writes: 'Radiesthetic phenomena are due to wave movement of a nature generally more complex than the transversal vibrations of light, in which the wave is associated with something, as photons are with light, which I shall call radion.' Certainly he may call them radion, just as he or anyone can 'call spirits from the vasty deep'; but the question is, as Hotspur queried, 'Will they come when you do call for them?' We need not pay too much attention to such pontifications. Just as I often feel no necessity to accept as poetry a piece of verse simply because it is written in blank prose, so I am unwilling to accept as profound truth a pronouncement simply because it is ill-stated. I am in the happy position here of not being obliged to make a statement myself on this matter – though I would have liked to do so. I do not think it is as yet subject to a complete statement. Meanwhile we can all console ourselves with the reflection that the divining goes on, the missing persons are found, the criminals are detected, the sick are comforted, the lovers are united, and fresh springs pour forth to replenish the thirsty.

3. The Primordial Powers

. . . We see water at the service of man as our life-saving drink, as a highway, as a force ensteamed by fire. But that is not all. Water harnessed simply to the force of gravity (a kind of fuel that will never become exhausted) serves us even better. In fact, we reserve the term 'power' for this alone.

All that is necessary is that falling water should hit a wheel with broad paddles (or vanes) placed to receive the blows, and it will revolve and go on revolving as long as there is a supply of water and a supply of gravity. Thus the invention of the water-wheel is the earliest and perhaps the most pleasing of all man's attempts to turn natural force to his advantage. And this, the first of his inventions, served the first of his needs – the necessity to grind corn. The mill soon took the place of the rough-and-ready quern stone. The old mill by the side of the stream in a sequestered nook in the quiet countryside – the very image of it fills the mind with thoughts of beauty and peace.

That was Act I of the business. As we have just seen in re-lation to canals, in the divine comedy of our drama on earth we are always obliged to press forward. Always the idea is: Cannot we do better than this? By better we generally mean faster. So here. We must make the wheel go round quicker. Summon the engineer. Thus the engineer came forward and suggested that instead of one or two paddles being struck by the falling water while the other paddles were idle until they came round into striking distance again they should *all* be struck at *once*. This could be achieved by surrounding the wheel and its paddles with a circular channel (with an entrance and exit hole for the water) which has a number of openings admitting the water exactly on to the paddles. Thus the water will hit all the paddles all the time and increase the velocity of the wheel according to the number of extra strokes – an immense increase. This is known as the turbine. In comparison with the old water-wheel, its power is terrific.[6]

6. How great it amounts to is well brought out in Paul Lewis's suggestive book *The Romance of Water Power*.

That was Act II, if I may put it that way. First the stream falling over the natural or contrived height and caught by the simple wheel. Then the elaborate wheel, the turbine, with its multiplied revolutions. And, I must add, with a multiplied power of falling water by virtue of the pipeline connected with the river or lake, from the nozzle of which the imprisoned water is jetted out with much greater force than when pulled by gravity alone.

This brings us to the third movement in this development. Great things were to be gained by this increased velocity. For now man held in his hand motion as a *tool*: not as a means of transport, not in the first instance, but as an instrument. Given motion to use he could call upon other forces to serve him. He could pluck forth from the endless spaces of the sky the hidden powers harboured therein, as one who might conjure spirits out of nothing. Laying hold of his instrument of motion he could attend to the mystery of magnetism and the genesis of electricity. No longer need the lightning flash forth only in the angry sky; he could conduct those forces from out their capacious dwelling into a logical habitation. He could generate power, and his generator is known as the dynamo.

It is beyond my brief, and my powers, to go into the workings of the dynamo. I would merely attend to the main facts before us. We know that there is electricity latent around us: we have only to stroke a cat's fur in the dark to realize this, just as we know that there is fire in the stones, since we have only to strike them together to see the sparks fly upward. We know also that we have the allied force of magnetism around us. So much is common knowledge. The scientists, the men capable of concentrating so closely upon the forces of Nature that they are able to work with them and even get service from them, contrive, by means of the dynamo, in which a magnetic field is juxtaposed with great frictional movement, to capture, to localize, to generate electric current.

The first task of the dynamo is to transform the power received from the fall of water into another power – that of electricity. The second task is to turn that power of electricity into other forms; for by itself it can only give us a shock. So we turn it back into motion again, making its greater power drive whole railway and tram companies; and we transform it into heat, so that we can have fires without need of fuel; and we transform it into light, so that whole cities and countrysides are inexhaustively illuminated.

See that lake high up in the mountains. It is a calm, restful, windless day. The lake shares that calm: it is silent, it is still. Yet – it is like a giant resting. It is energy asleep. It is a fuel, better than coal, better than oil for it cannot be used up; a fuel that creates power without perishing, and passes on unharmed, unburned, unwasted. The lake nestles quietly in the mountain basin. It is held firmly there in the arms of gravity: let any portion of it go over the side of the basin, and it will be pulled roaring down. In fact, over there we see a waterfall plunging into the valley. Then man comes – rather as if he were Nature's own right arm. The fall of the water is increased in force by the pipeline, the nozzle enters the turbine and the wheel revolves at colossal speed; the dynamo is fixed to the turbine and the speed is turned into electricity, and the electricity into the motion of engines and into heat and into light.

In making that summary we are surely stating the greatest practical achievement of man, without a single exception. Here he lays hold of the elements without abusing them – or himself by means of them. He robs no minerals from the exhaustible earth. He destroys no wood. He wastes no oil. Here is a fuel that is used, but not consumed, here is a force not exploded katabolically, but released metabolically in a smooth cycle; and as the cycle revolves he employs other forces that in a like manner are used, but not consumed, the primordial principalities and powers of the universe which fail not nor shrink, no matter how much service is demanded of them. We call upon the electromagnetic forces and rest upon their ever-juvenile, ever-replenished hoard. We may weary in our search for coal, we may cut each other's throats in our need of oilfields, but electricity will never weary. Our economy in this matter is the economy of Nature, and the great hydro-electric schemes the world over are rooted in the truth of unquenchable abundance.

CHAPTER VI

SECTIONS FROM '*SINS AGAINST WATER*'

1. The Corrupting of Our Waters

How can we sin against water? you may say. We can sin against
soil. We can sin against crops. We can sin against trees. But even
as the sun shines upon the evil and upon the good, so the rain
falls alike upon the just and the unjust, and as we cannot dim-
inish the rays of the sun, neither can we lessen the abundance of
the waters. That is true. Yet we can sin against water. We can
defile it. A river may be used as a water supply. It may be used
as a highway. It may be used as the means of power. But the
time came when it was put to a further service – it was used as a
sewer.

Consider. Once, and not so long ago either, all the rains on all
the earth were carried back whence they came along the river-
ways, taking with them only some mud and mineral properties
of the earth. There were no huge conglomerations of people
crowded into one place, and no industrial manufactures: the
rivers, the streams, the rills passed by uncontaminated. This
does not mean that people did not sometimes empty refuse into
them, or that many a corpse of animal or man may not have
found a grave there. In moderation this would not cause un-
healthy water. We know that when we put excrement into the
earth it does no harm; the soil deals with it and is even nourished
by it. It is just the same with water: impurities may be put into
it, but they are purified even as the soil purifies and makes corruption take on incorruptibility. For there is no such thing as
absolute corruption. There is composition, eternal becoming
and growth; and there is decomposition, eternal departing and
decay: but they mingle, they make one thing, a whole, and work
one upon another: any amount of decomposition cast upon the
earth will not corrupt the earth – this, as we all know, is the

mystery and marvellous miracle of compost. If there were such a thing as corruption in any final sense the world would be uninhabitable – nay, non-existent.

Now, it is no more surprising that water should act in this way than that soil should. The soil does it by virtue of bacterial activity – and so does water. There are as many microscopic animals inhabiting water as there are inhabiting soil. They have been extensively investigated and labelled under various heads such as Protozoa and Rotifera, Diatomaceae and Schizomycetes, Cynophyceae and Chlorophyceae. Since these big names do not help us much to envisage these small things, it is enlightening to study photo plates which magnify them 500 times, and thus be able to admire the variety of their composition: they call to mind so many things – a daisy, a cap with ribbons, a starfish; a bear with six legs, a moon with claws, a nut with horns, a bead with whiskers; a trumpet, a seal, a porcupine; a wine-glass, a shrimp, a curled-up elephant; a frog, a sponge, a worm; a wedding-ring, a bird, a surrealist painting; a tree-trunk, a crest, a flint instrument; a snow crystal, a bunch of grapes, a coronet; a fountain-pen, a string of beads, a cart-wheel without a rim; a fork, a leaf, a canoe, a coin.

In addition to these aquatic wanderers (plankton), though not wholly distinct from all of them, are bacteria responsible for the self-purification of water. 'When any kind of animal or vegetable refuse gets into water it is immediately attacked by bacteria,' writes H. D. Turing.

which act in rather the same way as fire though at a low temperature. The bacteria, in a sense, burn up the organic matter, using oxygen to do it and converting it into carbon dioxide, and leaving the residue as simple salts which fertilize the vegetable matter growing on the bottom of a river. This vegetable matter, in turn, under the influence of light, absorbs the carbon dioxide, fixes the carbon, which it uses for its own growth, and frees the oxygen, so the oxygen which was used by the bacteria is continually being restored to the water again.[1]

The main point here is that they use oxygen to do it. You

1. H. D. Turing, *River Pollution.*

cannot burn anything without oxygen, as we all know, for the quickest way of extinguishing fire is to smother it – that is, take away its food of air. Now, there is a great deal of air in water – it is said to be nearly half air, strangely enough. Water is composed of millions of round molecules like tiny marbles, a shape which necessitates millions of equally minute, or more minute, spaces, which are filled with gases – chief of which is air. The bacteria must have oxygen to do their work, and they get it from the water itself.

They are not the only water-creatures that need air. The insects and fish population also need it. If we fall into water, we drown because we do not get enough air. Fish can also drown through lack of air – if I may put it that way. It can have been used up. If a small quantity of refuse or organic matter is put into a river the bacteria will be able to deal with it without depleting the oxygen, and perhaps recover conditions within half a mile; but if too much has been put in they will fail to do so, the fish will seek other areas, the insects upon which they feed will die, the water become dead and no use to anyone.

Centuries rolled by and all the rivers of the world flowed to the sea uncontaminated. If we pass in review the entire known history of mankind from the beginning, we must say – we are bound to say – that not until the nineteenth century were rivers seriously *polluted* by man. They could not be. And now, in the twentieth century, many of them are so badly polluted that all life is being exterminated within them. That is where we have got to now.

How has this come about? In two ways: by domestic sewage and by industrial wastes. Before the advent of large towns, with their substitution of water-closets and the piping for the earth closet, there was no question of river-pollution on a large scale and it is interesting to note also that the phosphorous content excreted by the population was returned direct to the earth then, and not, as now, lost to an amount calculated as equivalent to 150,000 tons annually for England alone. This domestic waste can be, and often is, treated by a perfectly clear and efficient process of filtration and preliminary bacteriological burning. But often it is not treated, and is discharged straight into rivers as if they were handy ditches.

That is the first cause of pollution; the growth of towns causing a reckless discharge of diluted domestic sewage. The second cause, worse than the first, which has gathered momentum during the last half-century, is trade-waste effluents. There are a great many of them: there are those which consist of organic matter, rather like sewage, though stronger; or those which contain directly poisonous matter, such as 'acids, copper, cyanides, or which put into a river a considerable amount of solids in suspension which eventually settle out on the bottom and smother the insect life normally living there, or which contain a large amount of very fine matter in solution which interferes with light entering the water or with the breathing apparatus of insects or fish.'[2] Under the first head we may place effluents coming from textile manufactures and leather industries; from laundries and paper mills; from slaughter-houses; from milk factories and the wastes out of beer and spirits; and from a few others. Under the second head we may put effluents coming from coalmines and the coke industry; from steel and copper works; and from almost any kind of industry whose wastes contain metal.

Such are the sober facts. All those factories pass out wastes of a poisonous nature, and when they are not treated they have a devastating effect upon the rivers into which they are discharged, making them foul and dank and death-dealing to all aquatic life, the acids eating away the fins and rotting out the eyes of fish, while in some places the water has been so thickly dyed that it could be used as ink – as was in fact done with the Calder in Yorkshire when a page of a report was written by dipping a pen into the river.

The curse of water-corruption now faces all industrial and highly populated countries. It would be absurd to suggest that great rivers such as the Ohio or Mississippi could be polluted to a dangerous extent, since their power of self-purification must exceed anything which man could do to corrupt them; but

2. H. D. Turing, *River Pollution.*

Americans do not all live by the side of great rivers, but rather in river valleys or watersheds – such as the Brandywine River – where the flow is moderate enough to be easily subject to abuse. Yet no country has suffered so much as Great Britain in this respect. The consequences of pollution can best be illustrated by her example. Those results are known and tabulated. We are not in the realm of vague statement or mere angry exaggeration by anglers and fisheries. There have been many reports. And it happens that recently an Englishman, capable of that kind of selfless devotion and toil which saves nations without the nation being aware of it or of him, H. D. Turing, undertook the task of investigating from 1947 to 1949 all the most important rivers in Great Britain, making four reports thereon and a summary of the whole matter in a short book.[3] The four reports make a separate volume each, presented with a real sense of shape and clarity.[4]

It is only by studying such reports that we can visualize what is happening to our rivers in these modern days; how a river may receive contributions not in the form of lovely tributaries from lesser streams but from the poisonous wastes of fellmongers, tanners, paper-mills, calico-printing works, dye works, tar distillers, chemical works, sewage-disposal works, and gas works; how in fact, the very tributaries which it does receive may be utterly fouled; how for miles and miles on end rivers are thus rendered dead with no oxygen or aquatic plants or fish; how the combined contribution of untreated wastes often makes a river much dirtier than an effluent joining it from a well-conducted sewage-disposal works; how after the Industrial Revolution had set in with severity, the salmon catch on the Ribble went down between 1867 and 1914 from 15,000 to 280, while the trout on the Lune, a river which receives no less than ten sewage discharges, went down between 1922 and 1946 from 872 annually to 22; how the Avon, Shakespeare's Avon, is today a sewer for a large part of its course, receiving wastes from industrial towns; how the Torridge in Somerset at the

3. H. D. Turing, *River Pollution*.
4. *Pollution*, prepared for British Field Sports Society by H. D. Turing.

outskirts of Torrington received unpurified town sewage together with the water-washing waste from a milk factory at the rate of 3,000 gallons an hour; how paper pulp and other wastes form so thick a scum in parts of some rivers that birds find it solid enough to walk on without sinking; how in the Irwell and the Roch there are no fish, no insects, no weed, no algae or larvae, no life of any kind save sewage fungus, while the treated sewage effluents 'are hailed with delight as being the purest water which the rivers hold'; how the Afon Lwyd, the most poisonous stream in Wales, polluted by forges, rolling mills, ironworks, and collieries, has been deserted even by the rats on its banks; how hundreds of dead salmon (when any are left) are found in rivers defiled with poisonous matter, and fish with the edges of their fins burned away by caustic effluent, as in the Ribble, and nothing left of the eyes of codfish save empty sockets caused by the corrosive acids in the Wyre; how a small stream, the Isla in Scotland, only a few yards wide, receives 240,000 gallons of sewage (with sanitary paper floating on top) every single day; how the Jed on the Tweed watershed, at Jedbury, 'may fairly be described as a sink'; how some rivers have been made so sludged and smelly that uninstructed spectators have thought that they were not rivers but open sewers; how the once lovely mountain streams of South Wales, moorland brooks of purest water, fit for every human need, are today 'thick with coal dust, poisoned with waste acids from iron and tinplate works, contaminated with sewage and the waste products of the gas and coke industry, sometimes polluted in their extreme upper reaches with effluents from whale-oil and chemical works; lifeless, pestiferous, foul; the vomit of our boasted industrial development.'

The case of the estuaries is somewhat worse. Reckless as the discharge of untreated wastes has been into the rivers, it has been more so into the tidal reaches, the assumption being that salted water will be able to deal with anything. But it cannot possibly deal with tons of continual unloading which, before it can be carried out to sea, is carried back again by the incoming tide. The narrow, winding estuary of the Forth receives the entire sewage, untreated, of Stirling, which is churned up by each succeeding tide to cause a permanent flow of filthy water which in hot weather and at low water creates a nauseating stench along the riverside. The estuary of the Tees receives an average of 12,000,000 gallons of crude sewage every twenty-four hours, so that it is almost impossible for fish entering the river to reach their spawning-beds higher up. The long terribly defiled estuary of the Tow in Somerset is used for the discharge of fourteen sewers, only one of which is treated, while the Torridge receives six permanent discharges, together with a gasworks and its contribution of chemical pollution, the consequent foulness of the estuary making it necessary for the fishermen below Bideford to take their nets out to sea to wash away the excrement choking the meshes. The estuary of the Tyne is even worse, since Newcastle and Gateshead, and other large towns down to the river mouth, empty into the tidal reaches 30,000,000 gallons of entirely untreated sewage every day, which amounts to about 11,000 millions a year, with the result that at low tide human excrement can be seen dropping from the sewage outfalls – of which there are 268. The estuary of the Lune should interest holidaymakers, and I would not like to withhold the following advertisement issued by the Fishery Board for 1945:

A glance at a map of the north-west coast will show that the coastline between Liverpool and Barrow measures about 150 miles, into which is discharged a minimum of 200,000 gallons of crude sewage every mile every twenty-four hours, or about forty large buckets full every yard daily. This is the cesspool that thousands of people bathe in during the hot summer months, and this is where our shrimps and prawns live and feed.

If I close these citations at this point, it is not, I fear, because the list is anything like complete.

It is easy to see what this means to the fish population – especially salmon. We seldom see salmon in our shops nowadays. Salmon, as everyone is supposed to know, are fresh-water and sea-water fish. They spawn in the higher reaches of rivers and grow up there for two or three years, at which time we call them smolts – by a masterpiece of ugly nomenclature. Then they swim out to sea, and after three or four years they return to the river in which they were born. If they cannot get out through a polluted estuary or enter through the same, it will simply mean the extinction of salmon. It has been calculated that the net annual catch of salmon in the Severn has dropped from 24,000 in the early part of this century to approximately 3,000 now. In the Tyne, in 1872 it was 129,000. Between 1941 and 1945 the average was 701 (I like that 1), while rod fishermen caught only eight in 1941. In the Tees the average catch for the four years between 1909 and 1921 was approximately 8,000. In 1920 the Fishery Board were obliged to make the following announcement:

There is an almost incessant complaint now, both of net fishermen and rod-and-line fishermen, regarding the grievous pollution of the river and its tributaries. The pollution is causing the destruction of fish life, and is seriously prejudicing a valuable national food supply. Owing to the pollution in the lower reaches, the spawning fish entering the river find it almost impossible to make their way to the spawning beds, while the young salmon smolts perish in vast numbers when making their way to their natural feeding-grounds in the sea, thus seriously interfering with the future supply of salmon.

Supporting the cogency of this manifesto, it was found that in the four years between 1926 and 1929 the average catch had fallen from nearly 8,000 to under 3,000; between 1930 and 1934 it went down to 2,000; in 1937 the catch was 23; now it is 0.

2. Modern Water-purification

Such are the facts – or, rather, a few of them. Could such pollution be avoided? Yes. The remedies are known and practised when and where there is the will to practise them; and some industrialists are showing the utmost concern in tackling this affair. Even badly polluted rivers have been brought back to fair conditions by the application of those known remedies.

For the most part this is not yet done, for there is not the will to do it by too many of those responsible. It pays not to do it. Why not empty your refuse into your neighbour's garden if he is not strong enough to prevent you? The law of the jungle is never more evident than in this kind of thing – and with what moral indignation and effective laws does the strong man prevent the weak man from doing the same thing in a small way as he does himself in a big way.

There are a great many ways of being a scoundrel. There are a great many ways of deceiving yourself that you are not a scoundrel. The simplest is to pretend that you are a benefactor, like those industrialists who recently were kind enough to come forward and advise a committee that before legislation was enacted against pollution it should first be decided in the interests of the community whether it were 'better for the country as a whole that our rivers should provide pure water or whether they should be used as sewers'. Mere commentary upon the poverty of such minds and the gross materialism of such an outlook is of little use; what is needed is legislation with heavy financial penalty. Is there such legislation? No. Is there any legislation? Yes – there have been sixteen Acts of Parliament from 1847 to 1951. Do they work? No – they are too weak, they lack penalty, and they offer loopholes, all along the line. That is how things stand now in England: sixteen Acts of Parliament, and fifty-four reports, surveys, investigations, and scientific papers and books from 1865 to 1954. That is something. The next stage will be legislation which works. It is always thus. First the scandal; then the outcry; then evasion and false apologia and feeble restraint; then when things have got bad enough they get better – real action becomes imperative. 'Perhaps by the end of the next century', writes Mr F. T. K. Pentlow in summary at the end of his book (the latest in England on the subject), 'river pollution, except as an accident, will be a thing of the past'.[5] I quote those words, for I think there is reason in them; but rather unwillingly, for they will certainly not be true unless we choose to make them come true.

5. *River Purification.*

I suppose it will be done, if it is done, according to each nation's eccentricity. Yet I think we might take a hint from the Americans here. I like the Hoff method. Mr Hoff is the Executive Director of the Brandywine Valley Association. One of its most pressing problems at first was water-pollution from paper-mills. The equipment capable of treating the effluent from the factories called for an expenditure of $20,000. How persuade the mill-owners to produce it? Mr Hoff's method, as described by Lady Eve Balfour, is interesting. He would approach a mill manager with his colour camera. 'May I take a picture of the brilliantly coloured water of the stream?' 'Go ahead,' says the manager. Then, as an afterthought, 'What do you want it for?' 'Oh,' says Hoff, 'Publicity.' The manager is a little troubled. 'What are you going to do? Show folks what I am doing to the stream?' 'Could be. But I am hoping to show them what you are *going* to do to the stream.'[6]

Considerations of beauty will not influence our industrialists, but fear of publicity that might injure their pocket may make them less likely to oppose legislation. It should be demanded in the name of beauty as well as cleanliness. 'There is nothing more beautiful than one of our clear English streams flowing through lush meadows and pastures by tree-lined banks and past quiet hamlets to the sea, unless perhaps it is one of our Welsh or Scottish mountain streams, dancing and tumbling over the rocks to a music which is one of Nature's purest songs. Substitute for these – as has happened in hundreds of cases – receptacle for filth and garbage, green, yellow and brown scum, lifeless and nauseating, and you have a contrast which it is difficult to equal in any other medium.' Thus nobly and justly writes Mr John Rennie, Chairman of the Fisheries Committee and British Field Sports Society, in his Foreword to the Third Report by H. D. Turing. The reader who has been good enough to accompany me to the beginning of this chapter, and has gazed with me upon so much that is beautiful and strange, will not wish to evade these harsh facts which I have felt compelled to set down. We have come far indeed: from water as the most

6. *Mother Earth*, October 1953

powerful, and sometimes the most exquisite of what the mystic Jacob Boehme called types of the heavenly pomp; from water as the holy lotion which brought salvation; from water as the ablution that symbolized redemption from earthly stains – to water as a sewer. We need not call this progress.

I happened to pay a visit to the Natural History Museum in South Kensington just at the time when I was writing my chapter on water as purification. My mind had been moving in religious fields, and I had been thinking of the holy rivers and the sacred wells and the living waters and the rites through which men have sought to achieve purity and cleanse themselves from pollution. Having entered the building I walked across the hall and found myself in front of a big glass case. Behind the glass were various diagrams and maps and enlarged photo-plates of bacteria, the whole being labelled with two clear words – WATER-PURIFICATION. This brought me up sharply. This was our way. Here was the modern mode. In old times people thought of water as a means of purification; now our job is to purify water. It will not cleanse us from pollution: we must cleanse it from our pollution. And most of us, nearly all of us, live in places where we are very glad indeed to be assured that the water has been 'treated', purified, sterilized – made not poisonous. But is this Nature's gift? Is this a holy thing? Not quite, perhaps. It is not exactly the cup that inebriates. We are far away, so far away, from those Arabs and from many people still in many lands less progressive than our own, who offer water to their guests expecting them to comment upon its particular flavour, as if it were wine.

Those living amongst us today who believe in water as a drink with healing power are badly placed. It is hard for them to prove their case. It is obvious enough that pure water, real water, alone, with its incredible solvent power, should be able to solve many a physical problem, internal and external, and has done so for centuries. But few people can lay hold of it. They are forced to poke about in the medicinal atmosphere of 'watering-places'. Writing a letter from Clifton, Macaulay said: 'Carlyle is here, undergoing a water cure. I have not seen him yet, but his water-doctor said to S. the other day – "You wonder at his eccentric opinions and style. It is all stomach. I shall set him to rights. He will go away quite a different person." If he goes away writing common sense and good English, I shall declare myself a convert to Hydropathy.' (Carlyle's comment on Macaulay, by the way, when he once saw him in a club, sitting in a corner reading, is equally amusing, and rather more to the point.) We do not know whether Macaulay ever became a convert to hydropathy, nor how far the waters of Clifton humoured Carlyle's stomach or affected his style; but what a criticism it is of our civilization that people have to go to special places to get a drink of water that will do them good, and that their doing this has to be given a special term and dismissed as hydropathy or faith-healing.

What has always been true in the past is not necessarily false in the present. I know few things more encouraging than the accumulated evidence of those who claim – throughout the ages – the curative power of pure water. The importance of its mechanical function in the body is scarcely open to question. There is nothing cracked or far-fetched in the following:

In the first place it modifies the fluids in general, and the blood in particular. Secondly, it is a diluent and solvent of the solider alimentary matters and their waste constituents in the alimentary canal. Lastly, it is a vehicle for carrying the nutrient principles therein prepared into all parts of the organism. It serves also as a menstruum in which to dissolve and carry out of the frame its wasted, deleterious or diseased particles. In this last item of its mechanical utility, water is to be regarded as the grand purifier of the innumerable canals and reservoirs, pipes, sewers, and passages of man's complex machinery. This is no low or unworthy idea. It is an idea which has been advanced by some of the first names in science, as explaining the undoubted cleansing efficacy of the water cure in frames corrupted by the combined impurities of disease, diet and drugs.[7]

7. Dr. H. Wilson, *The Water Cure*, quoted by T. Hartley-Hennessy in her challenging *Healing by Water*, which is also something of an anthology of evidence of believers in water-healing, from Hippocrates to the present day.

It is not fair to expect medicine-men to press home this point, since they want to give us medicine, not water. According to Pliny, Rome was for 600 years without a physician. That was too few. Today we can hardly have too many, since their chief task is to salvage us from the effects of our civilization. It is very likely that the effect on our health, and teeth and every- thing else, of our not-drinking water is very great. For we do not drink water. We drink beer, spirits, tea, lemonade and all the rest of it. We have no desire to take water. Certainly I haven't. I do not feel drawn to chlorinated water – that is, water in which so much of that invisible organic life which exists in it and really composes it as a totality is exterminated. Everyone who is not in prison most of the day can take the sunshine. We must go up a mountain to find 'straight' water. It falls from over our heads, you may say. Yes, but we cannot get it, we cannot buy it. The flowers do very well on it. The animals ask nothing better. But we cannot have it; not even a bath of it[8] – which might work wonders for rheumatic people. We must be content with chlorinated water which we can boil and take (as I do) burning hot. That is the situation today. I do not exagger- ate. We stand on the earth. We look up. Water falls from the heavens – but we cannot have it. We look round on our rivers – and dare not risk drinking them before treatment. In the end we are extremely grateful to the sanitary authorities for supplying us with a fluid that will not actually poison us.

(Yet, following an account of the great success of the Ten- nessee Valley Authority, and then of the modern plans to re- claim the Sahara Desert, I concluded more hopefully).

Do we enter a new era in the history of mankind? In old times great civilizations have foundered, their cities going down in sand like ships in the sea. In modern times we have increased existing wastes and made new ones. But now a new idea has alighted upon the minds of men. 'Tell me,' said Monsieur Sau- magne, the Inspector-General of the Tunisian Administration, gazing across at the barren lands that were coming to life at Kasserine, 'tell me that there are ten men who believe that we

8. There were 30,000 public baths in ancient Baghdad.

can redeem the stupidities of mankind – and I shall die happy'. There are more than ten men ... Our story is not over yet. If there is folly, there is also wisdom and grandeur in man. Sometimes we realize this when we listen to a major symphony. 'What was making the little man so exultant?' wrote R. J. Cruikshank after watching Toscannini conducting. 'He was rejoicing with the creator spirit of the composer, proclaiming with him the dignity and splendour of the spirit of man, assuring us that all that was bad, and silly, and stupid, and cruel would pass, and that the glory of the Heavens would remain.' Already we begin to see man as no longer the demented clown who desecrates his own garden and wastes the bounty of Nature – but as Nature's own right arm. The great hydrologic wheel for ever turns upon its task. We can put a spoke in that wheel. We cannot break the cycle nor arrest the eternal return of Nature's vast relentless roll. We can suffer for it. We can pay the price in dire confusion. We can perish in the dust. Or, recognizing our follies and our sins, we can put our shoulder to the wheel.

BOOK III
THE TRIUMPH OF THE TREE
1950

My task has been much easier in this instance. I have decided to omit the whole of my Part on the Mythology of Trees. The subject is so delightful and I treated it at such length that I think it more satisfactory to omit all of it than to include bits of it. This decision gives me room to include most of the other parts of the book, though I have been obliged to cut out the Reply of the Trees in relation to Greece and Rome, and to the Empire Builders.

– JOHN STEWART COLLIS

ARGUMENT

We were nursed into life by trees. It is to trees that we owe the development of a physiology which made Man possible – that is to say, made conceptual thought possible. Those fundamental facts should be sufficient to explain the intimate quality of man's relationship with trees. There is a still more practical side to the connection. Trees are necessary to our existence because they are the chief guardians of the soil, keeping it stable and watered. In the very ancient past, trees were thought to be spirits or the habitation of spirits, both good and evil, and finally were conceived as simply deities who were the guardians of fertility. This climate of thought lasted for some centuries in every country and led to a very widespread worship of trees and to an equally widespread fear of injuring them. We call it the Era of Mythology. This way of thinking gradually broke up and we entered the Era of Economics when trees and everything else were valued in cash. At the height of this economic era the application of science showed how swiftly and completely men could make use of trees in particular and nature in general. We have just reached the end of that period, having found that such an attitude has brought us to the edge of disaster. We are about to enter what might be called the last act of the drama, when science now discovers precisely in what way trees really are the guardians of fertility after all. This will be the Era of Ecology – the science of achieving an equilibrium with the environment. Thus having come full circle, we are back at the beginning again. But is it too late to make a fresh start? The world is not what it was at the beginning of the story. Half the wealth has gone. Even so, we *could* save the situation. But are we sufficiently alarmed to mend our ways? Do not too many people think, on the contrary, that we have done so extremely well that we can now actually look forward to entering an era when we will experience freedom from fear and freedom from want? But it is unlikely that we will experience much freedom from want until we have restored our capacity to fear the responses of Nature.

PART ONE
IN THE FORESTS OF THE NIGHT

1. The First Plants

To make our way back to the beginning of trees we must cross
the boundless empires of Time, driving on over the geological
eras, unfolding and still unfolding the manuscripts of stone,
passing beyond the Quarternary and the Tertiary page, beyond
the Cretaceous and the Jurassic, beyond the Triassic and the
Permian, until we come to the measures of Carbon.

It is not easy to do this. The sheer length of time blurs our
conceptions. When we face the eons and behold their sculpture
our minds become dazed and we feel hopelessly separated from
the early scenes. We make an error here, I think. Time is real
enough; the carvings of its knife are proof of that; but we over-
estimate the importance of distance. We allow it to make things
unreal: even friends in a far country are only half-alive for us.
But that which is far away, in space or in time, is just as real as
that which is close at hand – and not nearly so unfamiliar as we
sometimes think. Happily we can overcome this error and dis-
perse this mist. For example, if we stand on the Scottish Highlands we can actually get a very good idea of what the earth was
like long before even the Carboniferous Age, before the
Devonian, the Silurian, or the Ordovician, before even the
Cambrian. From those Highlands we can actually *see* the pre-
Cambrian Age, we can stand upon the first rocks, the Lewislan
Gneiss. Gazing upon the dark unhuman peaks where no flowers
grow and no bird sings, we throw off the weight of time-distance
and see what could have been seen five hundred million years
ago. Certainly we can say without exaggeration if we gaze upon
the rocks and forget the verdant valleys, that the spectacle
which we see as we stand upon those desolate hills is much the
same, and perhaps in parts actually the same, as what could

have been seen at the beginning of earth's race when the mountains had been moved above the waters.

And if we wish to attend at the very beginning of vegetation we need not close our eyes and try to imagine the first clothing of the rocks when all the earth was bare. We can still watch the formation of the first plants. If we take a piece of bread and leave it exposed for a day or two it will become hoary with the grey stalks and powdery fructification of the mildew and the mould; if we throw a clean boulder into the glen beside its river it will soon be clothed with the first mercy of the moss; if we dam up water and make a stagnant pool, the liquid mirror will not long serve to reflect the face of Narcissus, for it will soon become as green as a lawn with the growth of confervae which in time will choke up the whole and turn it into a bog; if we destroy our cities and cast down our towers they will be covered soon or late by these humble workers of the dawn. Thus they toiled in the dawn of life. They toil today, we see them do it – but then they were all that could be seen of verdure. Then they grew flesh upon the stony skeleton of earth. Then they wove the tapestries of the hills. Then they were the highest forms of organic existence, the first step forward in the flowering; they were the kings of life, and as they spread their mantle over all the earth they laid the foundation for the kingdoms of the world.

2. The Primeval Ferns

It is likely that as soon as they began to advance at all they advanced swiftly, these moulds and mosses, lichens and algae, fungi and other cryptogamia. First a sort of slime no thicker than paint, gathering through the years in depth and texture and richness, laying a floor for tufty plants and at last a foundation for the first pedestals, the first *stalks*.

These last grew ever thicker and higher into forests of fern. Studying their print in the library of the rocks we are able to grasp the magnitude of their multiplication, their rank luxuriance, and the huge dimensions of their territory. In the library, one whole volume – vol. VI of the Series, coming between the Devonian and the Cambrian – is devoted to these plants, an era called the Carboniferous, calculated as three hundred million years ago, and as lasting for uncertain millions of years. This chapter in the history of the world, according to some authorities, saw this vegetation spread over the whole earth, from Melville Island in the extreme north to the Islands of the Antarctic Ocean in the extreme south.

Again we can see what was there to some extent by looking at their descendants here. We know what bracken looks like when fully grown – a miniature forest. Then that scarce penetrable phalanx of reeds and bracken attained the height of forest trees in our day – as if the horse-tails of our bogs were magnified sixty or a hundred times, as if our reeds, mosses and ferns had sprung up as giants. Larger still were the various kinds of tree-ferns, the Lepidodendrons and Sigillaria, called club-mosses, though they much more resembled palms and cycads. Their stems were ribbed and fluted like Gothic columns, and they grew leaves thickly, tier upon tier, gradually shedding the lowest. And mingled with them were some genuine trees, conifers such as we might recognize today – the first real trees to appear on earth. They were not proportionally large as were the ferns compared with our growths, not giants to our dwarfs, but normal.

These forests grew in very swampy soil if we can call it soil at all, with a carpet of fallen leaves, spore-dust, prostrate stems, mosses, and liverworts at their feet. It would have made squelchy sinking ground for any creature to tread upon who might wish to roam there or penetrate into the mystery of the measureless bogs and stems which then made up the landscape of the earth. No creatures did attempt this, for none existed. At least no air-breathing animals at first. There were plenty of water-breathing animals. Motor existence had advanced well beyond its earliest forms in the marine world. Some had run their race, giving way to others. The sun was sinking upon the kingdom of the trilobites, the horseshoe crabs, and the eurypterids, while lobsters, hermit crabs, and true crabs took the stage, and scorpions, with twelve eyes disposed in a circle, entered the scene. But the air was not yet breatheable. It had four times the amount of carbon as in the whole of the atmosphere today. The tree-ferns, in their enormous abundance, purified the air.

These tree-ferns, writes Fabre,

constituted the greater part of the gloomy forests that were never enlivened by the songs of birds, nor resounded to the trample of the quadruped. As yet the dry land had no inhabitants. The atmosphere must have been unbreatheable, for it contained, in suspension, in the shape of poisonous gas, the enormous amount of carbon which has since become coal. But the tree-ferns, like other plants of their day, set to work in order to cleanse it and to render the solid earth habitable. They subtracted the world's carbon from the air, storing it in their leaves and stems; then falling into decay they made room for others, and these yet again for others, which unremittingly pursued, in the silence of the woods, their noble work of atmospheric salubrity. The purification of the atmosphere was at last accomplished, and the tree-ferns died. Their remains, buried underground, have in course of time become coal measures, in which leaves and stems, wonderfully preserved as to form, are today to be found in abundance, and record, in their archives, the history of this ancient vegetation, which has given us an atmosphere that we can breathe, and has stored up for us in the bowels of the earth those strata of coal which are the wealth of nations.

The appalling silence, from day to day and from year to year and from century to century, must indeed have been savage. Could we stand in such an atmosphere for a few minutes it would fall upon us with the oppression of impending calamity. Better far, we would feel, to hear the wild cries of pain that are the price of life, the shrieks and howls that answer the gift of air. We do not know for how long the unspeakable sadness of that silence bent over the flowerless beeless seas of green, but the air must have been breatheable for some centuries at the close of the Carboniferous Age, for the existence of many insects is established in the archives of stone. Here the spiders began to set up their everlasting line; the leaf-eating worms appeared in company with bugs, fireflies, lice, snails, centipedes, crickets, and cockroaches, while the flight of dragonflies, with a wing-spread of twenty-inches, is vouched for by the photographers of clay. And gradually the Labyrinthodonts, precursors of the Dinosaurs, ancestral newts, whose long bodies were protected by an armour of plates like the tiles on a roof, emerged from the water and developed lungs. They have left us their foot-prints in the sand. They took a walk, three hundred million years ago – and we, today, know all about it!

3. The Sinking Forests

How did these forests come to write their story in rock and to measure their weight in coal? 'Falling into decay they made room for others, and then yet again for others.' They did that, and they did more. They sank. At intervals they were over-whelmed by flood, and the land surface of the forest disap-peared. In due course sediment settled upon that submerged vegetation until it appeared above water again and served as the basis for a new forest. (That is one school of geological theory; another is that the submerged land was pressed upwards from time to time by forces below.) Whichever it was, the process was repeated. It was repeated numerous times. In Nova Scotia up to a hundred forests were buried, one above the other, with the roots of trees found still in their original positions, and with some of the trunks still standing erect; while at Essen the rocks reveal to us that *one hundred and forty-five separate forests*, each a land-surface on its own, were lowered down one on top of the other.

Thus, after drinking the sunshine these forests went down into the bowels of the earth to a depth sometimes equal to the height of Mont Blanc. And then, by cause of the terrific weight upon them, by the temperature, and by the pressure of the enor-mously patient, unbreakable and unbending tool of Time, they were squashed into pulp and hardened into rock. We call it coal.

A bed one yard thick takes a thousand years to make – thus three million years were necessary to produce the South Wales field.

We take a piece of it in our hands, a black stone. It is carbon, it is sunshine shaped into a solid. It is a piece of the sun itself we hold, the blazing ball itself turned into the dirty darkness of that rock. It may be very cold, freezing to the touch on a winter's day; yet still it is the ancient furnace that we finger, it is heat made cold, a frozen burning beam. We do not doubt this for a moment. We know how to change it back, how to make it into fire again. We put a piece of its own element in touch with it – its own essence, flame – and in a few minutes the box flies open and the trebly millioned years' imprisoned sun streams out, and the ransomed rays that fell upon the ferns fall on us today.

All the plants of those forests were not ferns. Some of them, as already mentioned, were conifers such as we might recognize today – the great group of Gymnosperms. Thus here, in these carboniferous vistas, is the cradle; here the first nursery of trees. In these glades was matured the idea of not falling down.

Nature advances by process of big new ideas. Fall down and get up again – that was the old idea. Get up by virtue of seeds thrown down. Most vegetation still does this. The building of most plants collapses utterly after a season, and a new representative has to be set up. Then the great New Idea – to have stalks that would not wilt, to have stems that would not die, to set up columns that would outlast pedestals of stone! Thus trees. We can see the unfolding of the idea among the ferns. A tree-fern shows us the beginning. Having set up a pedestal for its fronds, the fern seems to have said one day, Why not start from here next time, instead of from the ground? A good idea, but how carry it out? If they were to have pedestals which would remain fresh and also be equal to the burden of supporting a growing capital, then certain novel architectural devices must be employed. In due course they were devised, and anyone who is prepared to visit any fairly comprehensive Tropical Garden Museum (such as Kew Gardens in London) is likely to see what could have been seen in those early days – ferns starting to set up their pedestals, and ferns with pedestals already established. Thus arrived the club-mosses, the sigillaria, the palms. And once the New Idea had been liberated and exhibited in visible form in the outer world, it was taken up and improved upon by other species, the first of which were conifers. When we hear the wind blowing through the pine trees, the least imaginative of us is stirred; we are carried back into the depths of Time when that melancholy moan was sounded long before the trampling of any feet.

4. The Lapse of Ages and the Yeast of Creation

The scroll of Time unrolled. Calculating the incalculable, and endeavouring to measure our minds against the crushing milestones, we name the period that followed the Mesozoic, and count it as two hundred million years duration, ending some eighty million years ago. Once we grasp the fact that Time is so formidable, so utterly outside our ordinary conceptions, we can then see it as a concrete thing, rather like a chisel or a knife, at work upon the surface of the world, changing the shapes beyond recognition. The Palaeozoic and Azoic periods that preceded the Mesozoic are said now to claim about fourteen hundred million years. If we can face the impact of that Force we can accept without surprise the advance in organization as exemplified for instance between the slime-mould which parted company with some of its own members when meeting obstruction and later reassembled them (the parts being hardly more than jelly), and the gigantosaurus which exhibited a body a hundred feet long. Given that commodity, Time, in unlimited abundance, the creative force of life could do what it liked. The eye that could pass in review the periods that followed the earliest forests would behold the emergence into the now purified air, life, no longer stationary like trees, but moving freely on never less than four and not more than four hundred supports or legs. Beneath the waters many other animals could have been seen at this time and well before it; but now they rose from the deep and in different shapes took to the air and the land. We have some of them with us to this day such as the tortoise, the crocodile and the lizard. The interesting thing is that at this early period in the history of animals the largest organizations appeared. The Dinosaurs are so far the top size in animals. That limit was never surpassed. And, as everyone knows, they did not do so well as the smaller animals, and they perished utterly. Perhaps their eggs were eaten by the smaller animals. But there is no great mystery: they could not possibly have survived a severe climatic change. They were succeeded by creatures who did a new thing.

Another new idea. Animals evolved who gave up creating their young by laying them on the earth packed in a hard o or oval. They kept these o's or ovals inside themselves and presently delivered their young ready-hatched. And when they were hatched their mothers took an interest in them and gave them milk. These mammals, as we call them, set in motion a form of life never since abandoned by succeeding generations. From that time thousands of new forms became possible and life multiplied in such staggering variety that it looks as if variation is boundless and life as much a unity as water.

The great hammer strokes of Time continued to fall upon the flora and fauna, and upon the appearance of the earth itself. Continents rose from the deep and sank again. The earth we know today, was not. Our Africas, Americas, Australias, Indias had no place, they were more and they were less. The vast Gondwaland that today we see no sight of, for a brief space of a few million years lay exposed to the sun where leagues of waves wash now. Lowlands became highlands and highlands were cast down. Endless eras lapsed. The fingers traced upon the stones are silent in relation to huge gaps of time; but we do know that during all this time, while life spread out more and more into ever new formations, while the creature about the size of a fox changed into what we call a horse, while the giraffe pushed up that neck and the elephant unrolled that trunk and so on, forests, ever increasing in strength, dominated the lowlands of the world. Not everywhere perhaps; but the amount of unbroken forest stuns the imagination, it appals the mind. Picture the space of an Atlantic Ocean all trees. Some of the forests cannot have been less than that – land oceans of leaf as seen from above.

And down below, under the surface of that leaf-ocean, what went on? There boiled the cauldron of life. There swarmed the yeast of creation. In those depths were prepared the fiercest and furthest forms issuing from the womb of the mighty mother. The silence of the carboniferous days was broken by the snarls coming from carnivorous jaws, by the clamour of growing life, by the moan and bitter burden of being born. We need not seek in imagination that dark backward and abysm of time to conjure the scene: it can still be watched in certain parts of the world, such as in the Bolivian jungles. There the traveller can still behold the frightful ferment of primary creation. 'I felt carried back twenty million of years to the primeval world,' wrote Arthur Heye after having made an expedition to a pool in the Brazilian jungle. Speaking of the atmosphere at night, he says:

Between the trees the turbid water was rippling gently beneath the fiery red of the blazing moon; the melancholy call of birds sounded in the gloaming, an almost inaudible rustle of wind went whispering through leaf and reed. But then the water was stirred with weird motion: a gurgling, splashing and rustling, an ever wilder spattering and plashing. The whole of the dead water was heaving with invisible life. And gradually I began to understand what life this was – and my hair stood on end. Those were crocodiles. The whole pool seemed to consist only of crocodiles! And now at night the reptiles were apparently falling upon each other; a continuous hollow bawl and roar arose from the centre of the pool; a furious lashing of tails, a loud rattling of jaws, a foaming bubbling of the water; and over the crest of the surging waves flashed the dusky red reflexes of the spectral moon.

He stood at a narrow branch of the water and saw no less than five hundred crocodiles pass by: they must have lain one on top of the other in the lagoon like packed herrings. It was the very yeast of creation: 'This is what the nights of the Jurassic sea must have been like, when the fights of the ichthyosaurians were raging amidst the vapours of carbonic acid clouds, and the lurid light of the still fiery moon gleamed down upon the· scene.'

Again, the Spanish writer, José Eustasio Rivera, in *La Vorágine*, speaks of the jungle with the same sense of early days – the overwhelming battle to come into existence and to remain in it, the destruction and renewal, the scent of death, and the ferment of procreation.

Here dwell the responses of bloated toads; here are the pent waters round rotting reeds. The aphrodisiac parasite is the master here, strewing the earth with dead bees; here is the varied wealth of obscene flowers contracting like sexual organs, whose sticky odour inebriates like a drug; here is the malignant liana whose downy beard blinds the animals; and the pringamosa which enflames the skin.

And his description of the night scenes would surely stand for the unwitnessed spectacles of the primal years when the law was writ large that all coming into existence is fraught with filth and horror. What the philosophy of pure Spirit, as Keyserling put it, would banish to the very depths of Hell, is the earthly womb of all life.

In the night unknown voices, phantasmagoric lights, funereal silence. Death passes on its way and gives life. There is the sound of fruit crashing down with the promise of seed as it bursts; the fall of the leaf filling the mountain recesses with vague sighs, the offering of itself as dung to the parent tree; the crunching of jaws eating for fear of being eaten; the squealing of the disturbed, the moans of the dying, the belching of creatures easing themselves. And when the burning dawn reveals the tragic splendour above the mountain peaks, the tumult of the surviving sets in: the cooing of doves, the grunting of boars, the grotesque laughter of monkeys.

5. Trees the Prerequisite of
Conceptual Thought and the Birth of Man

This brings us to the next step of the unfolding. Amongst the new Appearances in the forest there rose the monkeys. They did a new thing.

Hitherto there had been the earth-bound animals with legs and the air-borne animals with wings. The monkeys left the ground without riding the air. They made the trees their habitation. In due course this promoted the development of something new in the history of all the creations. The monkeys started with four legs in the ordinary manner. Gradually they changed their two front legs into something else – they created *arms*. And while they discarded front legs for the new things called arms, they also substituted something else for the front feet – they made *hands*.

The most important thing which ever happened in the forest was the creation of the hand.

The Arm, the Hand, the Finger, the Thumb: consider the destinies that lay enfolded there!

This brings us to Man.

For the next great event was the appearance of the first men. How did they come? We all know the popular theory about this. When the lemurs or apes or monkeys had after thousands of years so altered their bodies as to stand upright on two legs and to possess real arms and hands – then the First man appeared. Presumably he looked just like the Last ape of that line – with an infinitesimal difference which multiplied and multiplied until real men such as we could recognize today in a primitive community became common, and in their turn increased.

Such is the famous story of man's evolution from the animals. It has not yet been determined whether we really stemmed from the apes, or whether we had a stem of our own from which the apes degenerated. It is unlikely that any Missing Link could clear this up. And the main mystery does not seem subject to solution. When we talk about evolution we are concerned with two things, one easy and the other difficult. It is easy to grasp and accept the main idea of evolution – that of an endless variety of species forming by the clash of growth with circumstance. It is distinctly queer that one of the animals should have ceased to be an animal and become something else. Why did only man do what man did? All the other animals went on being animals for thousands of years and seem likely to continue to do so. What is the difference between being an animal and being a man? It lies in the fact that man possesses conceptual thought and the specific instrument that goes with it – language. How is it that no other animal has broken into conception and speech? The answer is perhaps given in Dr Julian Huxley's very illuminating essay, *The Uniqueness of Man* by his detailed assertion that *only* the physiology of the Primates (lemurs, apes, monkeys) could promote the evolution of conceptual thought. That is, one animal did not cease to be an animal, but became an animal with conceptual thought. Still, we are left wondering why, if this is so, no monkeys are ever found now slowly becoming men. And if monkeys are de-evolutionary types of men, why do they not show signs of real speech?

Whatever the solution may be, man did arrive. And we must note the responsibility of trees in this matter. The existence of trees was a prerequisite of conceptual thought. It was the tree that gave the hand to life. It was the tree that promoted the upright posture. It was the tree that 'laid the foundation both for the fuller definition of objects by conceptual thought and for the fuller control of them by tools and machines.'[1]

Must we not say that it was the tree that gave man to life? Anyway, it was under the auspices of those ancient boughs that he appeared. And having appeared, he was destined to interfere with the Order of Nature. The very forests were doomed.

We are accustomed to think of nature and man as the norm. But man has only been at work for a million years out of the

1. Julian Huxley, *The Uniqueness of Man*.

two hundred million (if that is the figure). During all those years nature developed according to the terrifying Order which we can still see. Had it continued thus for the last million years, imagine the world today. Consider a journey through it. Not a man, not a city, not a machine, not a road: an almost universal forest broken here and there by bare mountain peaks and the great waters. Think of the surging life breaking ceaselessly like waves upon the shore and as ceaselessly dragged screaming back into the jaws of death: the untamed beauty of it all – unseen, unfeared, unpraised. Then man enters. He looks like any other organized piece of Nature. Will not the world of life go on just as it has always done? But he soon showed that he was utterly different from *all* the other stationary and motor organisms on earth. His eyes communicated something to his mind which the eyes of the other creatures did not communicate. He saw the world objectively. He became a spectator – something quite new. We call it consciousness. The other animals had eyes, ears, hearts and brains, but not this, and no tendency to evolve this.

That is the curious thing. It holds most interest for us today. In the last century it was man's similarity to animals, his kinship with them, his ascent from them, that came as a revelation. Now that we have grasped the reality of evolution one way or another, it is no longer our likeness to the animals that interests us and stirs us most deeply: it is our unlikeness to them, our singular gift of consciousness that has separated us from the flow of nature. Man alone stood back and looked at the world. He stooped down and picked up pieces of it, and used those pieces – even fire. The centuries roll on and yet no other creature is ever seen using a tool or lighting a fire! That wonderful Hand, given us by the tree, began to change the appearance of the world. Up till then life was in the hands of God: now man became one of the hands of God – nay, God was delivered into the hands of men.

6. Man's Cradle in the Forests

We will probably never know much about the earliest dates of man's existence. Some anthropologists speak of his arrival as occurring 'after the first Ice Age'. This is very difficult to believe. The line of his descent (or ascent) was obviously arboreal. His existence presupposes the existence of forests, and forests presuppose anything but an ice age. It is much more likely that man appeared before the first Ice Age and then taking to caves survived into the next warm period. The geologists give us four such cold periods. It is extremely hard to visualize them. These ages are spoken of as if something like a world-wide glacier slowly expanded over the earth. At the same time we hear of men living in caves and going out to hunt a variety of animals, and after hunting sitting by fires. Where did they get the fuel? Where were those animals and how did they exist if it was an ice age? We must suppose that these ice ages were not so severe as they are made out to be.

Happily this is not my concern. It is my privilege to unwind the ball of Time until we come to primitive man in the primeval forests slowly working his way into consciousness. He began, we began, amongst trees. Our story began when the earth's surface consisted chiefly of water and wood. The oceans! the forests! – the very words stir us still. They carry us back into the wondrous depths of the dawn, echoing and re-echoing our first fearful fumblings in the dark.

We began amongst trees. Can we imagine now what it was like living in the days of that beginning? We can have some idea; for though the corrosions of Time do carve incredible changes on the creatures and the landscapes of the earth, there are some things which repeat themselves so closely as seeming to ignore the passage of centuries. Amongst such we may consider the tropical forests of today. If we know something of them we can imagine something of the environment of our earliest ancestors, and therefore something of their feelings.

They came into the first glimmerings of consciousness in the forests. That was their world, all they saw, all they knew – a world of trees. If ever they were in a position to see the tops of the trees they would see a green prairie stretching out ocean-ically to the ringed horizon. From below the forest could be seen as a kind of unlimited table with tree-trunks for legs. To this day many animals inhabit these table-tops as if they were the earth below. Squirrels and sloths, tree-porcupines and tree-ant-eaters, tree-frogs and toads, not to mention a variety of birds, inhabit the forest-roofs without necessarily ever descending to the ground – so that even tadpoles are born in the high leaves that have gathered water, and there, in the arboreal ponds, suffer the stages of metamorphosis. We might truly call these surfaces *table-lands* in which the creatures live: not flying in air, not swimming in water, not stationed on soil or rock – but slung between heaven and earth.

Such a canopy above meant much gloom and darkness below. And it was there that men must have dwelt in forest clearings. Dig a hole in the Atlantic Ocean; make a clearing in the water, and tent yourself down there in that cliffed-in place – and you probably would not feel more overwhelmed by the walls of water than in the old days when we lived at the bottom of the seas of trees.

When today we enter the tropical forests we do so with perhaps a million years' consciousness of the world behind us. And whether we live in it as natives or visit it as travellers, we are still subdued by it. It is not easy for a modern European to grasp what it means to dwell in a primeval forest. As I write these lines I sit within the shade of an English wood. I hear the hum of wasps and bees, and the call of birds, and in the distance the voices of children (whose cries echo far where there are trees). It is not wholly quiet nor silent. Yet quietness is here, and silence. Nature's holy nuns surround me in the congregation of their convent. Here I am set in the midst of liberty and peace. In these gentle climes trees are friends to us, and in their presence, amidst the calm of their cathedral aisles, we are prone to meditate upon the blessedness of life.

It is not so in the great forests of the world. There we will not find benediction. The silence is as terrifying as the noise. The oceans of unvoyaged verdure, the brutal entanglement of parasitic lianas that rope with wild riggings the branches of the endless trees, make a scene that soon tames and humbles down the hardest man. The spells of silence, silence us – nay, subdue us utterly by sense of the violence pent up in that stillness. Then it explodes. It cracks across. Thunder goes off like a cannon behind our ears. Earth's foundations falter. Rain and hail fall as if an army from above were hurling down stones and buckets of water. Fiery bolts of metal crash like bombs. The darkness is figured by screens and serpents of white fire. This is the end, we cry. It is finished. All is lost. The world reels back now to its beginnings. Then the tempest stops as suddenly as it began, and the dawn breaks upon earth still fresh and strong, glinting in the rays of the rising sun. Our fear is cast aside and our hearts expand in joyfulness and praise.

Thus even today we know Fear in the forests. Otherwise we seldom know it. A thunderstorm is good fun, we rejoice in it. The wilder it is the better, we are glad to see nature violent and scathing. But we have lost that fear which is a clutch at the heart, and in order to achieve it we must resort to the thunder of oncoming enemy aeroplanes. And now some people are advocating freedom from fear. That might be all right from a physical point of view, but how about the metaphysical? In the Middle Ages three monks came to Father Sicoes and complained that they were continually pursued by three things: fear of the river of fire, of the worm which dies not and of the outer darkness. When the Saint made no reply they were greatly distressed. Finally he said: 'My brothers, I envy you. As long as such thoughts live in your souls it will be impossible for you to commit a sin.' We must not forget that religion unfolds from the womb of fear. The most intense form of religion is superstition. The most superstitious man is the most religious of all religious men – for he is most in touch with, most sensitive to the *numinous*, the Other, the invisibilities, the god-forces. It is not at all a good thing to be too religious, for then you will far too often be foolish and frantic; but I do not know how it will be proved that to have no respect for the elemental powers, even while using them to blow yourself up, is to be wiser and more profound.

Let us beat our way back across the deserts of Time and think of man coming into consciousness as he dwelt in the deep forests of that ancient day.

7. The Roots of Mythology

'Coming into consciousness.' A phrase. We write it down. We take it as thought and pass on. But what's the good? Let us think what we are saying. To come into consciousness: can we, who are in consciousness, imagine this event which changed the history of practically all life on earth except some insects, and which altered the top surface of most of the world?[2] The animal walked along, it still walks along, submerged in the stream of life in which it moves, though it is not like a drifting piece of wood, since it has an engine of its own and is conscious of itself. But it is without consciousness of the world as something it can look at objectively. How do I know this, seeing that I am not an animal and have never looked out at the world from behind the pale windows of that kingdom? Because if it had conceptions we would soon know it. I look at the sun rising. An animal looks at the sun rising. It sees the same thing. But it cannot see it as the sun rising. It cannot see it as something to be thought about and named. No animal can do this nor seems ever likely to do so however long its race survives or however much it may evolve – though unless our stem is a unique growth, one animal once did so.

Anyway, the thing happened. One creature broke away from the determined life, drew back, looked on, and perceived the world as something to be conceived. Can we imagine now what it felt like to be the New creature emerging into this awareness?

2. Every anthropologist will realize that here I telescope the process: some thousands of years may well have passed before man's consciousness became differentiated materially from the animals, and became *self-consciousness*.

We know what he felt like physically: the movements of the legs, the stomach, the throat, we know what it felt like to be him in that way. Very well; now let us take the eye. He saw out of it just as we do. But what did he see? Supposing for the sake of argument that the change was sudden, what was the difference between the last day when he was an ape and the first day when he was a man? Here we cannot think from experience. We do not remember. All we can say is that on that last day objects must have stood or moved in front of his eyes unidentified as objects, though they might react upon his senses in terms of fear or hunger. On that first day there was a communication exchanged between his eye and his brain, and an object was picked out, *separated*, and perceived as an object in itself. It is a pity we do not know the first word he spoke when he broke into speech. It must have been a sound that meant a question. By asking that question and picking out that object he took the first step in separating himself from the flow of phenomena.

Having become aware of the world, he at once set to work to separate everything in it, to sort out one thing from another. He did this by *naming* them. He began giving names to things, and today we will not find anything in the world that has not been given a name; or, if such a thing is found it is regarded as a great discovery. He began to give names to men themselves, so that today it is impossible to find a person without a name. *What is his name?* we ask, assuming that he has a name as certainly as he has a head or hand. The gulf between ourselves and the other creatures is seen most clearly in matters such as this. They live unnamed amongst each other – such is their lack of conception. Think of a formicary of ants, each ant with a name: impossible! – even amongst monkeys impossible.

Having become aware of objects and begun to name them, this Earliest Man became aware of something else. It is a remarkable fact that no sooner had he looked closely at the phenomena of Nature than he began to concern himself with, not the visible object in front of him which he could clearly see, but with an invisible object which he could not see at all. He looked at the trees, the rocks, the rivers, the animals, and having looked at them, he at once began to talk about something *in them* which he had never seen and never heard of. This thing inside the objective appearance was called a god. No one forced man at this time to think about gods, there was no tradition imposing it upon him – and yet his first thoughts seem to have turned towards a Thing behind the thing, a Force behind or within the appearance. Thus *worship* – the very thought of which in connection with any animal seems ludicrous.

All this is such an old story and has been gone over so often by anthropologists, this evolution into worship and birth of priesthood, that we tend to regard it all as quite natural. After the event things always seem natural and obvious. But before the event? One wonders how many of the heavy-weight anthropologists could ever have prophesied the birth of worship as an early result of the birth of consciousness.

To sum up: suddenly (if in fact slowly) the circle was broken, and certain creatures, that is to say ourselves, achieved the power to look at the world with surprise and question. The motion of the wind, the utter transformation of trees in season, the coming and the going of the sun, the fall of water, were observed as strange things. The dawning mind began to ask questions. Surrounded on all sides by the extraordinary nature of the ordinary, faced with the strange object of a tree, and having fixed it down with a name, what did we do? Did we then look at it calmly and examine it? No! – that is the scientific spirit, something not to be born for many centuries. We immediately ceased to look at it, to see what was there. We saw what was not there. We obscured its existence with an orgy of fancies.[3] We saw beings that did not exist, we heard voices that spoke not, we bowed before innumerable spirits whom we named. This remarkable attitude of mind has of course been dealt with by specialists the world over. It is now called Mythology and has been examined, docketed, labelled, categorized, and shelved.

3. Yet behind the fancies was the intuition of living unity that we call *animism* – which is far from fanciful.

PART TWO
THE TRANSITION

... Then the gods died. Pan perished. Later he was to return as Pantheism: but in the meanwhile he and all his crew were withdrawn from their earthly tenements and held no more traffic amongst men.

The old conception of plurality became worn out, being at last too much of a strain upon belief. The idea began to gain ground that there was only one god and that he was not attached to the earth, and neither subject to view nor interview. This new conception had the vitality of freshness and might have served as well as the old idea if all the races of mankind could have agreed upon the identity of this single god. Unfortunately there were disputes over him, one race projecting one favourite and another claiming quite a different person – each candidate being conceived as extremely jealous of his god-head. The resulting clashes between them make by no means the happiest chapters in the history of religion. It is a cheering thought that at least we can claim to have reached better days on this score, now that it is recognized that neither monotheism nor polytheism touch the core of religion, which is a mystic experience of the Divine.

What concerns us here in the change from the first idea to the second, is that whereas under polytheism the gods were intimately connected with the earth, and stimulated veneration for it, under monotheism deity was extracted from the earth. God was promoted to higher regions. He was completely out of sight. It became possible to fear God without fearing Nature – nay, to love God (whatever that meant) and to hate his creations. This attitude reached its climax with what is called Puritanism. In the history of mankind there is nothing more shameful than the spectacle of human beings perspiring with religious fervour and at the same time turning away with horror and loathing from a fresh green leaf or a naked body.

That frenzy passed. Then the Deity began to be secretly hated. But this did not mean the restoration of the gods and a renewal of respect for the creation. Science came in and began to 'conquer' nature. The gods were dead, and God was dead: into what channel now would energy flow? Into the exploitation of the earth. All respect for every living thing was abandoned; the Golden Bough was turned into boughs of gold; the once god-informed trees became 'timber'; and the way was made straight for the princes of industry and the kings of commerce.

The above is merely a statement of the transition. The history of the transition would be obliged to take into account the untidy nature of change. The gods did not die all at once, nor quickly. It took time to get rid of them. In vain did the Hebrew Prophets denounce the groves; in vain did Mahomet put tree-priests to death for honouring acacias. The worship died hard. The ecclesiastical records of the Councils of Agde, Auxerre and Nantes were forced to make prohibitions against sacred trees and woods, and to treat as insane those who burnt candles in honour of trees. These records, says John O'Neill, 'give us an all-powerful motive for the almost cosmic crime of the fatal destruction of the European forests.' That and cupidity. The usual mixture of motives. If trees militate against the worship of the God above, destroy them; if trees are not venerable then use them for money-making.

It will be my task now to consider how we stand today after some centuries of the new attitude towards the earth introduced by monotheism and carried forward by commercialism allied with science. Sorry if this part is rather short. I see no necessity to write a hundred pages marking the transition – always supposing that I possessed the qualifications for doing so. But no candid person, with some knowledge of the past and with his eyes open today, will deny the broad truth of my statement. And what concerns us is not the process of change but the fact, and our present position as a result of it.

The result has been bad, as we shall see – and indeed as everyone knows. But if we retrieve the situation we shall be in a happier position than we have ever been before. Tree-worship, as we have seen, evolved from crude forms into the general idea that they were the guardians of fertility. There was a profound truth in that. Unfortunately the truth was seen in a false light. The trees were thought to be alive in the wrong way – to be gods or the habitation of gods. Thus when people no longer believed in these deities, the trees ceased to call for veneration or care. We are gainers, if having passed through that stage, we now enter an era when respect for the earth is being reborn, and science, no longer speaking in terms of conquest, can inform us with a deeper truth why trees really are the guardians of fertility.

PART THREE
THE REVENGES OF NATURE

CHAPTER I

TREES AND THE NATURAL ORDER

1. The Equilibrium of the Primal Order

Before going a step further let us take a firm grasp of the obvious and remind ourselves of something too easily forgotten. We must remember that man is a late arrival on the scene of natural operations. It is not his fault that he has come; he did not ask to be born; and his birth is the strangest thing that ever happened. It was so unnatural. There seems to be inherent wisdom in the workings of Nature. In this instance that wisdom is hard to discern. It is as if a bird had laid a bomb instead of an egg.

No one has been able to explain this. Let us not try. Rather let us contemplate for a few minutes the Order of Nature without reference to man. Is there an Order, is there really any Natural Order? Well, the opposite to order is chaos, and the universe is remarkably coherent. There is government throughout.

We have just passed through a short phase when the public has been slightly bewildered by scientific men who happened to be rather weak thinkers, amusing themselves by comparing organisms with engines as if there were no difference between a pistil and a piston; by confusing description with explanation; by declaring that the universe is a chaos of conflicting atoms; and by failing to observe that when they insisted that everything is governed by chance they had thereby acknowledged that it is – *governed*. The idea of Purpose or lack of it was introduced into all this. Grave questions were asked and elaborate volumes written on this question of purpose, as if the movements of molecules could determine it. What was really being asked was – Is there any point in life? This question can only be answered emotionally. The universe is given and makes a total impression: we value it according to our psychology. Any man is entitled to say that he hates it. For myself, I like it, or a lot of it. My appreciation of it cannot be undermined by any description of its operations nor by the suggestion that in so many million years the sun may cool and go out. I applaud its existence whether it lasts for ever or comes to an end one day, whether it is run by atoms or by a god on a throne somewhere. A purely personal reaction, subject to my psychology. But what is not subject to psychology, what also cannot be altered by any talk about 'conflicting atoms', is the fact that the finished article, the universe, is unified, is governed, is lawful.

What are we saying when we speak of the Order of Nature? We are saying that the unnumbered phenomena that we see existing in an intricate complex maintain a marvellous degree of order and balance. They melt into one another. They are one another. A cow standing there in the field with its huge wet nose, looks singular enough. But where does the substance of that heavy piece of flesh it carries, come from? The cow is not created out of nothing, it is not really on its own, it is simply part of the universe, a moving part. There is the field; it happens to be stationary. The cow moves about, but it is part of the field, and is continually recruited from the field. It cannot move, it cannot grow unless it takes in a portion of the field. This is called eating. If it fails to do so, then it will stop in its path, and will sink back into the field. This is called dying. Now, it is in a high degree undesirable, and useless to fall into the pit of speculation and ask. Who contrived this ? or to explain it in any way: these are questions we must always pass over while doing our best to describe it and fully realize it. Having done so, having laid hold on the obvious before our eyes, we can pass to less visible types of the heavenly pomp. Take the case of the fish, the snail, and the plant for instance.

Place a bowl of water where the sun can strike it, and put a snail, a fish, and a water-plant into it. The populations of three in this cosmos will thrive for months by virtue of mutual exchange. The fish lives on the plant. The waste of the fish is prepared by the snail so that it can be manufactured by the plant which uses the waste of both fish and snail for its own purposes, and in so doing releases oxygen that purifies the water and guards the animals from suffocating. When the physical and chemical equilibrium is thus maintained by the action of different creatures, we get what ecologists call a balanced environment. Remove the snail, and the plant will droop and the fish will fail. Remove the fish, and there can be no further exchange between the plant and the snail. Remove the plant, and neither fish nor snail can pasture. It looks like three in one and one in three. But of course it is more than that. We must not forget the water. We must not forget the sun. Take their action away from the fish, the plant, and the snail, and again this cosmos will totter to its foundations.

What are we saying by virtue of the above illustration? Only the obvious – that when things are in their right place they are all right. But are they not inevitably in their right place? The Order of Nature either is an order or it is not, and everything must be in its proper environment. Nothing could ever start in an improper environment, otherwise it would never start: the bloc only breaks into many organisms when the conditions are favourable (at least such is my vision of the matter). Thus when we speak of the Order of Nature are we not saying something as obvious as that the sun rises and sets with marked punctuality? Yes, I think that that is really all we are saying – that the universe works, that it does not break down. There is an inherent principle at work which ensures a balance.

I am happy to make that elephantine truism. There is a fiction that people fight shy of original ideas. It is supposed that you cannot get original ideas into people's heads. This is untrue – we love a new idea to play with. What we cannot get into our heads is a platitude, a truism. Let us get this one in – that nature does preserve an equilibrium. And then let us accept the fact that this does not rule out *calamity* for individual units or even whole species if the Order at any time is overbalanced. Occasional calamity of individuals is part of the Order.

Thus we do not find one species becoming predominant over the rest. It is not possible for any species to achieve this. About three-quarters of a million species of insects have already been identified and therefore named.[1] A single pair of plant lice, it is said, could produce enough progeny to outweigh the human population of the world fivefold. In six months flies can raise six trillion offspring. We have all seen what a plague of caterpillars can do to a wood within a month. We have all heard what happens when a cloud of locusts appears. The Australians know a little about rabbits under this head. The white ants in massed millions have sometimes more power than an earthquake. How is it that a balance is preserved between all these creatures, and that one lot does not empire it above all the others? This is what all units attempt to do. But when they succeed too well their food gives out and the conditions in which they normally thrive are changed. In Nature nothing fails like excess. And normally the balance is held by virtue of each succumbing to enemies which keep them down. Thus birds pay insects great attention. It is estimated that in the U.S.A. three billion breeding birds are insectivorous. The proud cockroach whose line is linked with ancestors before the first forests of the world, and whose digestive system can easily assimilate 'a mustard plaster, an Egyptian mummy, or Jefferson's *Manual of the Constitution*',[2] is itself readily digested by a variety of birds. During the fledgling period the starling feeds insects to its young at the rate of up to three hundred and fifty visits a day, while the ordinary house-wren brings back an insect every two minutes. It gives us an insight into the resistance which nature offers these tiny creatures when we learn that the potato-beetle is attacked by twenty-five species of birds, the alfalfa weevil by forty-five, the coddling-moth by thirty-six, the gipsy-moth by forty-six, house-

1. According to Macneile Dixon 'there are not less than ten million varieties of insect'.

2. See Stuart Chase, *Rich Land, Poor Land*.

flies by forty-nine, bill-bugs by sixty-seven, cut-worms by ninety-eight, leaf-hoppers by one hundred and twenty, and wire-worms by one hundred and sixty-eight. Nor are birds their only enemies: the bat and the shrew, the mole and the squirrel, the badger and the skunk are also fond of insects. But should these rodents thrive too well thereon, they in turn are looked after by the birds. The various owls like nothing better than rats, mice, shrews, voles, and rabbits. When in 1907 a plague of field-mice invaded the region of the Humboldt River in Nevada, gulls, hawks, and owls came on the scene, and in one month had disposed of one hundred thousand mice.

It is difficult to visualize the amount of creatures, big or small, that exist in the world. Could one member of each species pass before our eyes as in parade it would take some time before the show was over. Would we then, gasping at such numbers and such variety, wonder how they could all get on together? Or is that the right word? Have they developed the faculty of living together? When the 100,000 mice fell before the gulls, the hawks, and the owls, was that not failure to live in unity? In speaking of the Order of Nature, are we not merely exchanging a happier phrase for 'Nature red in tooth and claw'? When, under days of frightful frost and ice, there is a huge, silent massacre of birds, what right have we to speak of Nature as well ordered from the point of view of the birds? To this we must reply, I think, that the redness of Nature's tooth and claw is part of the order; and that the word Order cannot be equated with Painlessness, but rather with the Basis-for-existence. Again we are bound to acknowledge that our reaction to existence on such a basis, is a matter of temperament. We should invite the birds to take a metaphysical view of the matter and to remind themselves that only appearances are against them, and to sing with Emerson: 'If the red slayer thinks he slays, And if the slain thinks he is slain. They know not well the subtle ways I keep and pass and turn again.' We are foolish if we try to do without philosophy in this world. We should always bear in mind the words of that unprofessional philosopher, Anton Chekov:

So long as a man likes the splashing of a fish he is a poet. But when he knows that the splashing is nothing but the chase of the weak by the strong, he is a thinker; but when he does not understand what sense there is in the chase, or what use in the equilibrium which results from destruction, he is becoming silly and dull as he was when a child. And the more he knows and thinks the sillier he becomes.

Silly? Can Chekov be allowed to sweep aside heavy-weight philosophers in a single phrase? I don't see why not. Yet I would hesitate to use the word myself. For there are passages even in Macneile Dixon's tremendous book *The Human Situation* which might thus come under that head. He writes:

How difficult to recognize in the ferocities we see around us the subtle power which made the brain, which elaborated with consummate exactness the mechanism of the heart and lungs, all the devices by which the body maintains its existence! That Nature should create a world full of difficulties and dangers, and thereupon proceed to place within it fabrics of an infinite delicacy and complexity to meet these very dangers and difficulties is a contradiction that baffles the understanding. With a cunning past all human thought she solves the problems she has, as it were, absent-mindedly set herself. The flood and the earthquake have no consideration for the plant or animal, yet Nature which sends the flood and earthquake has provided, with foresight or in a dream, for the living things they destroy. She both smiles and frowns upon her own creation, and is at once friendly and unfriendly. Like a scarlet thread it runs through her dominion, this inconsistency. Side by side with the undeniable and admirable adjustment between things organic and inorganic, you have the hostility, the discordance. What wonder that men, bewildered by this inexplicable procedure, have supposed her governments distributed amongst a hierarchy of squabbling deities, persecuting or protecting this or that race of men – Zeus for the Greeks, Jehovah the Jews. What wonder they supposed even the trees to be the better of protecting deities, the olive Athena, the vine Dionysus? Ah, Nature! subtle beyond all human subtlety, enigmatic, profound life-giver and life-destroyer, nourishing mother and assassin, inspirer of all that is best and most beautiful, of all that is most hideous and forbidding.

Thus the author of that great book – *The Human Situation* – himself extracts a deity, as it were, an outside organizer, calling it 'she' or 'Nature'. Need we make that approach? How would it be if we do not postulate an outside organizer of the whole in that manner – but take as given an inexplicable growth? We would be no worse off. Dixon is not pleased with our tendency to speak in terms of unity and wholeness. 'That the world is a unity,' he continues, 'the philosophers and men of science reiterate with wearisome persistence. That it is united they have the sense not to proclaim. How the world became disunited they have not told us,' Certainly these are words to make us think again. Yet what is he saying? That the units are not linked in fraternal embrace – not that they are disunited. But everything goes to show that they are linked, they do form a whole. Yet what of it, it may be asked, if that whole is not to our liking? it is a fair question. For there is nothing necessarily meritorious in unity if that article is unpleasant. We are entitled to reject it utterly if we consider that there is too much room for improvement. But on the whole men do not reject it. Quite the contrary. And Dixon, who always faces all the facts, is careful to acknowledge this before closing his chapter.

Our immediate concern here is with the equilibrium which results from the complex of laws and forces. The reader knows that the few illustrations which I have given above, are as a thimbleful out of the lake of known facts concerning the interactions and pleasing exchanges of natural life, and that the known facts are as a thimbleful out of the ocean of what is still unknown. But we have enough to go on, we know that there is a balance, that order is held finally amongst phenomena however often and no matter upon what scale temporary disorder may break out. The Order of Nature should be a more fruitful conception than the Survival of the Fittest. The latter phrase – contrary to Darwin's intention – conveys merely that those survive who do survive. Up till very recently it has been assumed that many offered proof that the strongest could survive – not matter how much he expanded, or how much he disturbed the scheme with his tools and his guns and his insecticides and all the rest of it. That was the famous 'conquest of nature' – which is beginning to look a bit thin now.

2. *The Rule of Return*

We have just glanced at the best-known and seldom forgotten aspect of the wheel of life – the way in which each creature receives others into itself and gives itself to others. Man appreciates one side of this anyway – the receiving of others into himself; though he so rarely holds to his side of the bargain by giving himself to others that it is regarded as an event when anyone is eaten.

This brings us to another very important aspect of the Order – namely, the ever-active movement of return. All these creatures of the earth whose numbers could never be counted, are, as it were, motor crops. They come out of the earth, and when they die they do so by either entering into the bowels of others or into the bowels of the earth – in both cases giving further vitality to the devourer. They come into the world and they receive the gifts of the world, its air and water and soil: this is called growing. In due course they give it all back, all of it. Some ecologists say that they give back a fraction more than they receive. This may be so. If it is, then the world is adding to itself, increasing itself. Yet this seems unlikely. I do not see how there can be such a thing as total addition or multiplication. It is more likely that there has always been and will always be the same amount of substance: yesterday the world a flaming ball, today changed into earth, air, and water: nothing added, nothing subtracted, all conserved. A question of change, not of addition. Indeed I have little doubt that the learned could point out to me that if the totality could grow as its units grow, we would very soon notice it; and I suppose that the law of the Conservation of Energy carries in itself a denial of total Addition. So I beg leave to question the soundness of any ecologist who suggests that when any living creature dies and renders its account, it is in a position to give more than it has received.

The main thing, however, is that it does return what it has received, not only in terms of a corpse, but also when it is alive. When it receives another creature into itself, when it eats that creature, it can only use a portion of it and must pass out from its body the remainder. We call it waste or excreta. But of course waste is an absolute misnomer. It is no more waste than a corpse, and can be devoured by certain other creatures and by the earth itself. The creatures in the natural order do not think of it as waste, they do not think of it at all; they just return it haphazard to the earth. This goes on day after day, year after year, century after century, countless tons of excreta being returned to the earth. To the earth: not to the sky, nor to the sea. The amount of this, for one single day, if assembled in a heap would stagger the beholder.

This unbroken circuit of return of good received, provides one of the most glaring aspects of the Order. And how we step outside it! I am not ready yet to speak at any length of Man in relation to the Natural Scheme; but it is proper to remind ourselves at this point of what is perhaps the most civilized of all things concerning him. He does not hesitate to receive into himself as many creatures as possible (no animal in fact approaches him in this), and he also can use only part of them, and must chemically treat the remainder and pass it on. But towards this matter passed out he has an attitude all his own. Wherever he dwells there is found a private little place called a *closet* or w.c. by means of which he can keep the product almost invisible from himself, and by means of pipes can keep it totally invisible from others. These pipes and underground passages, which he calls his sewerage system, are designed so as to prevent any of this product, which he calls waste, from reaching the soil. It reaches the sea and the rivers: 'in England we waste every year 219,000 tons of nitrogen, 55,000 tons of phosphate and 55,000 tons of potash as sewage sludge and house refuse that pollute the rivers and are lost in the sea.'[3] Europe and the United States combined, dispose of 20 million tons of nitrogen, potassium, and phosphorus every year in this way. The stuff is

3. H. J. Massingham, *The Tree of Life*.

not lost, of course; it remains in the world but in the ocean – that is in the wrong place for the good of earth.

This subject is painful indeed; we are right to draw back from it – for the quality of what we pass out, its smell, is such a terrible comment upon our physical corruption in comparison with the animals. We are right to feel ashamed and wish to hide our product, when we witness the bowel action of a horse: the easy delivery, the sudden ending, and the clean flesh at the finish ... What a far cry it is from the coral reefs which are the excreta of polyps, to the sewers of London and Paris! How strange that the first should proceed from the lowliest of all the children of the world, and the second from ourselves who claim to be the highest.

3. The Natural Creation and Renewal of Soil

We have looked from our window at one aspect of the dance – the circuit of the creatures. But the wheel includes the earth upon which they dwell: the balance is also between air, water, soil, and plant.

The earth was once chiefly rock. If we wish to face the foundation of life and gaze upon the mystery of the first floor, we must look at the rock. It seems steady. It appears lasting. But it is passing away. It cannot face the scrutiny of Time. For it is being weathered by wind and water, and Time is but the tool of Motion which is eternal. Fractions of stone fell, and falling laid the foundations of a soil. If that one poor figure is felt to be inadequate to describe the history of nature covering many millions of years, we cannot help it – perhaps the bare fact is good enough.

A rocky pile in modern times may take a long time to vanish; but there were periods of acceleration of the process by means of volcanoes, earthquakes, ice ages, and excessive violence of storm. The fact remains that under the rule of movement a portion of the rocks was turned into a substance ready to receive the first forms of verdue. I shall permit myself to write *turn into* verdure, since it seems more sensible to say that than to say it received something which was not there. A few more million years roll on and we see the mass of carboniferous forests. Everyone knows, whether vaguely or in detail, the rest of the story. We have just glanced at the Order of Nature governing the life and death of the creatures who dwelt and dwell among the plants. How about the soil itself?

Let us rehearse the well-known facts in a brief paragraph. Since soil starts as rock it naturally contains mineral properties. I need not enumerate them here, but we should remember that nitrogen, phosphorus, and potassium are three of the most important since they make the growth of plants possible. In fact plants rose from them, or they turned themselves into plants, and anyway still feed plants. Airy substances! Most delicate and light! We can see how easily they could be washed away. How complex is this stuff we call soil! An amazing integration of those chemical properties, of the rock particles, of the plant ashes, of insects amounting to millions per acre, of quite countless bacteria – no wonder we cannot make an inch of this material ourselves! We call it simply – humus. Obviously it is highly vulnerable. But in the Natural Order it does not suffer harm. The tempest breaks out, the wind whips the earth, the torrent falls; but the foundations of the house of life are neither blown nor washed away. For the soil has fostered those who foster it. It is nailed down by grass. It is pegged down by trees.

This does not mean that there is no destruction. There is destruction. There is natural destruction – or erosion. That is to say there is renewal. The scientific term is denudation. The process is described with such admirable clarity by Mr G. V. Jacks that it would be a pity not to quote his words:

What is usually known as 'geological erosion' or 'denudation' is a universal phenomenon which through thousands of years has carved the earth into its present shape. Denudation is an early and important process in soil formation, whereby the original rock material is continually broken down and sorted out by wind and water until it becomes suitable for colonization by plants. Plants, by the binding effect of their roots, by the protection they afford against rain and wind and by the fertility they impart to the soil, bring denudation almost to a standstill. Everyone must have compared the rugged and irregular shape of bare mountain peaks where denudation is still active with the smooth and harmonious curves of slopes that have long been protected by a mantle of vegetation. Nevertheless, some slight denudation is always occurring. As each superficial film of plant-covered soil becomes exhausted it is removed by rain or wind, to be deposited mainly in the rivers and the sea, and a corresponding thin layer of new soil forms by slow weathering of the underlying rock. The earth is continually discarding its old, worn-out skin and renewing a sheath of soil from the dead rock beneath. In this way an equilibrium is reached between denudation and soil formation so that, unless the equilibrium is disturbed, a mature soil preserves a more or less constant depth and character indefinitely. The depth is sometimes only a few inches, occasionally several feet, but within it lies the whole capacity of the earth to produce life. Below that thin layer comprising the delicate organism known as soil is a planet as lifeless as the moon.[4]

The only difficulty about the above is in the last sentence. Personally I am no believer in these vast gaps. Still, it doesn't matter here. We need not fall into the pit of more-than-physical speculation. We are not dealing with ultimates, at least not with ultimate ultimates. We may proceed. Soil, then, is the resultant mulch from the decomposition of rocks effected by the action of gravity, by the force of heat and light, by the play of air and water and ice, and then composed into a further complex by the rotten leavings of the very growths which rise therefrom. And further, these plants use their excreta (for plants do excrete) to break down the rocks below. The rocks above have been broken by weathering; the floor below is further quarried by roots. It is always so with Nature – the strong things are weak, the weak strong. The mountains are brought low, we tread upon them, by motions in the airy waste; and then the excreting juice from roots has power to crack the stones beneath that soil and thus increase it. These are the mills of God that grind so slow but do grind exceeding sure. If we would understand the Order we must forget the clocks of men and attend to the time-

4. Jacks and Whyte, *Vanishing Lands.*

sheets of nature. Soil-making is not a speedy process in our eyes: the rate is something like an inch of soil created every five hundred to a thousand years – or one-tenth of an inch in a century.

4. The Wheel of Water

Loam soil is said to be composed of one-quarter water, one-quarter air, and one tenth organic matter – 'it thus swims, breathes and is alive'. So say the authorities. Whatever the exact proportion of water in soils may be, we know that it is very great, and that in fact nothing in all nature is more important than water. The reader knows this. He also knows that this book is about trees, and he has been kind enough to accompany me this far in the confidence that I will stick to my subject. That confidence is not misplaced. I learn what not to do, as well as what to do from my masters. If John Ruskin had handled this theme, he would undoubtedly at this point have done a volume on water. But I am all for sticking to one thing at a time. This is difficult of course, since *inter-relatedness* is the very definition of Nature. The full theme of water is a great subject and can lead us into many places and unveil some surprising spectacles; and perhaps if I get back to trees the reader will join me at some future time in a consideration of water.[5] But something must be said here and now about the hydrologic cycle since trees play an important part in the smooth working of its flow. Incidentally, it is also another pronounced example of what we are calling the Order of Nature. My best plan will be to make a general statement, eschewing all detail and all the fascinating by-paths of the subject. Then bearing it in mind, taking the water-wheel as given, we can return to our trees and examine the specific part they play in the universal scheme.

There is a given amount of water in the world, residing in ocean, air, and earth. The total is neither added to nor subtracted from; it is constant. Though constant in amount, it is in

5. My book, *The Moving Waters*, followed later.

continual motion, passing in a ceaseless cycle from earth to ocean and ocean to air and air to earth. Its existence is necessary to all living things, and its *even flow* the prime factor in relation to their comfort and their quantity. Happily, over very large areas of the earth's surface the give and take is consistent and steady. The sun beats down upon the ocean, the ocean raises evaporation into the form of clouds which, like vessels, are shifted by the wind across the land. As the clouds pass over they are either obstructed or cooled and consequently fall to the ground in the form of rain. This is called precipitation. Having fallen, it may, if the land is bare, quickly evaporate again and rise back into the sky; or it may flow in torrents down a mountainside into the valley below; or it may strike a forest and neither evaporate nor rush down the hill, but slowly infiltrate into the soil. It depends where it falls, but for the most part we may say that there is a satisfactory exchange between ocean, air, and earth so that the fall of water from the sky is neither wholly sent back to the sky nor quickly conducted to the sea again. It enters the earth, thereby supplying immediate life-giving elixir to plants and animals, and also conserving the land against the days of drought in the form of lakes and reservoirs above and below the surface – for the amount of reservoirs and rivers and lakes below the surface of the land is nearly as great as above it, while the subterranean wells and springs are of the first importance in the economy of Nature.

The hydrologic cycle therefore presents us with another aspect of the primal order; and if the exchange between precipitation, infiltration, and evaporation is evenly held, then the best conditions for living things are provided. Nevertheless my description above is incomplete even as a bare general statement. For there is a complication which makes a neat statement of the cycle or the circle, impossible. In fact it makes the words cycle or circle a little dubious. I am thinking of the evaporation from vegetation. There is the evaporation from the ocean, and of rain sent up again from bare land under heat. That is one kind. There is another kind. There is the water sent up, invisibly sprayed into the sky by vegetation itself. A full-grown willow can transpire up to five thousand gallons in a single summer day. How much then a forest? Clouds can be made that way over the land, without benefit of seas. These are tree-clouds, not ocean-clouds. There is a cycle all right, and a constant amount of water in perpetual circulation. But we must not forget this cloud-feeding by plants.

No more of this at the moment. We are not likely to forget it, for our task now is to go into the theme of the ministry of forests in the government of nature.

5. The Influence of Trees upon Temperature, Rainfall, and Swamps

We have seen how vulnerable the soil is. It could so easily be tossed about by the elements, if unprotected. Happily it is protected, in the natural state, either by a carpet of grass whose network of deeply diving roots holds it down firmly, or by trees that on a tremendous scale stake it down, the tree-roots ramifying in all directions, so that only a hundred trees occupying an area of five miles will be actually supplying, in sum, three or four miles-worth of cordage for holding the soil together. Thus when trees were regarded by the uninstructed minds of superstitious men as the guardians of fertility there was some sense in it. When they become simply 'timber' in the eyes of unsuperstitious and instructed men who cut them down indiscriminately the consequences are so bad that modern science is busy restoring the idea that after all trees do guard the fertility of the soil.

Let us now plunge into the centre of our subject. On closer examination we find that trees perform many more offices in relation to the soil than that of merely pegging it down. By virtue of cooling the air and spraying the sky and multiplying the clouds they exert considerable influence upon the fall and distribution of rain; by virtue of sponging the earth around their feet they enormously influence the behaviour of floods, the discipline of rivers, the supply of springs, the health of fish, and (when man arrives) the welfare of navigation; and by virtue of their power to suck up moisture by the ton they dry the swamps and control the malarial mosquitoes. Forests are so much more than meets the eye. They are fountains. They are oceans. They are pipes. They are dams. Their work ramifies through the whole economy of nature.

The rays of the sun beat down upon a barren place. The naked earth becomes very hot and the temperature of the air very high. But if vegetation covers that ground the temperature will be altered. It will be considerably cooled. For the vegetation will evaporate water. It has been proved that, in terms of corn, for every pound of dry substance produced there is an evaporation of two hundred and thirty-five pounds of water; and in terms of turnips, for every pound of substance nine hundred and ten pounds of water is sent up. Under good cultivation an acre can produce seven tons of dry substance. On these terms we can calculate that a given acre will easily evaporate, during the vegetative period, about three thousand five hundred tons of water which will mount upwards to moisten and cool the regions of the sky.

If this is true of crops, how much more does it apply to forests. And further, we must remember that leaves do not become heated nearly as easily as rocks or open soil, while the ground under the shade of trees can never be greatly warmed. The result is that forests exert a moderating influence on temperature. That great mountain, the Brahmaputra, has not many trees; but its middle part is covered by forest – and there the temperature is less than at the bare parts by twenty degrees! The largest forest area in the world is at the upper Amazon, six hundred and twenty miles from the Atlantic on one side, and cut off from the Pacific by high mountains. So far from seas, so near the equator – will not the temperature be very high and very dry? Yet no, it is not greater than at the coast, and not as high as some temperatures in the middle latitudes. This remarkably moderate temperature is attributed to the enormous transpiration of water from plants in the tropics. The rainfall-down is about sixty inches a year. The rainfall-up (or evaporation) amounts to forty inches a year. Between the two lots the air is considerably cooled.

And in a perfectly ordinary way, the influence of trees upon temperature is obvious to every woodman. Not only in the way of cooling the air, but of warming it. Let us have William Cobbett on this from his treatise on Woodlands. Here he is – commas and all:

A coppice is *always* warm. In the coldest days that we know, when hail and sleet cut your face, and when you are really pinched with the cold, go into a coppice, and you are warm. In the very hardest frosts, the *ground is seldom frozen in, or near, the middle of* a large and well-set coppice of six year's growth or upwards. Even in that bleak and terribly cold country, New Brunswick, where the frost comes about the 7th of November, freezes the River St John (a mile across) over in one night, so that men walk across in the morning; where, in the open fields, the frost goes four feet down into the solid ground; even in that country, if you, in the very coldest weather, when, in the open air, you dare not venture *ten yards* without protecting your hands and face with fur; even there and then, if you go half a mile into the woods, you are in a mild and pleasant climate. I have, scores of times, gone to the edge of woods, wrapped up in flannels and blankets and furs, and when I got in, reduced my dress very nearly to an English one, and set to squirrel hunting, even with my gloves off.

This capacity of trees to moderate the temperature besides being so agreeable, is also a factor bearing upon the quantity and distribution of rainfall. Water comes down from the sky in the form of rain or snow or hail, and is further found as dew, hoar-frost, and other condensations of moisture which form on the surface of foliage, branches and trunks. And before it comes down, as everyone knows, it is tanked in clouds, or as clouds. What induces the exchange? Cold obstruction. That is why mountains promote precipitation. But wooded mountains are still more effective in deflating the fleeting vapours. Denuded hills do not always induce rainfall, while tree-clothed hills do. Dr Paul Schrieber, a noted meteorologist, after giving elaborate data for Saxony, reached the conclusion that in a district completely covered by forests the influence of the forest in increasing rainfall would be equal to elevating the region six hundred and fifty feet. We cannot easily raise mountains when we wish to increase the rainfall. It is therefore worth realizing that by judiciously planting trees we can lever-up a mountain about six hundred feet.

This cold obstruction induces greater condensation in the air-currents – and hence precipitation. But also, forests, whether on mountains or not, add to the weight of clouds by the evaporation we have been speaking of. And since they add to their weight they induce their downfall. The amount of water evaporated, that is, thrown off by forests into the air, is so enormous that they have been given the name of 'the oceans of the continent'.

These oceans go up into the air and then come down again. The subject is rather bewildering. It is not made easier by the authorities. There are always those who insist that the case is not proven, and that our knowledge concerning the relation between trees and rainfall is meagre and tentative, the hydrographic methods of measurement resting upon anything but exhaustive hydrometrical data. Over against them are those who give elaborate instances checked by experiments carried out through a series of years. My information is drawn chiefly from Mr R. Zon's *Forests and Water*. At the end of his report he includes a bibliography of other pamphlets and reports on the subject, amounting to 1010. I have not consulted them all yet. But I have a measure of confidence in Mr Zon, and certainly subscribe myself the humble and obedient servant of him and the other devoted men who strive to bring light to this subject. True, the instrument of language on occasion tends to break down in his hands as when he fails to differentiate with sufficient clarity the words *consumption* and *absorption*, or *evaporation* and *transpiration*; and there are times when he appears to be speaking of transpiration under the head of precipitation.

For it should be clear that when we speak of trees increasing rainfall – on the ruling of the above data – we are not talking about precipitation so much as of transpiration; we are not talking of rainfall-down so much as rainfall-up; we are not saying that the rain falls from the sky and waters the trees, but that it rises from the trees and waters the sky. Mr Zon makes some remarkable observations. He declares, drawing upon further authority, that seven-ninths of rain is supplied by land-evaporation, even over areas adjoining the ocean. He maintains that only seven per cent of all the water evaporated from the oceans enters into precipitation over land. 'It may be assumed therefore that the moisture which is carried by the winds into the interior of vast continents, thousands of miles from the ocean, is almost exclusively due to continental vapour, and not to evaporation from the ocean.' And again, he declares that 'seventy-eight per cent of all the precipitation that falls over the peripheral land area *is furnished by this area itself.*'

That is equivalent to saying that when we stand in an area of forest in the interior of a continent, receiving rain upon our heads, it is primarily proceeding from the ground beneath our feet. If you don't like this, note the following from another honoured authority, Mr H. S. Person, in his report, *Little Waters,* sponsored by Roosevelt: 'Depending upon regional climatic conditions, a given store of water which has been blown in over the land from the ocean in the form of clouds, may be "worked" three to five times as rainfall, because of alterations of evaporation and transpiration with precipitation, before it returns to the ocean as stream flow.' That rather supports my bold image above. Five times the sky has emptied a shower of rain on my head, and five times it has been sent back again before making its way home by river. It also supports my earlier statement that there are tree-clouds as well as ocean-clouds. The trees are fountains that invisibly spray the heavens with their exhaust. See yonder cloud hanging above the wood. There is a strong wind blowing, but the cloud stays at the same spot. It does not move on. The wind has no effect upon it. It remains above the those trees, hovering there like a hawk waiting for its prey. It should have gone miles away by now horsed upon that sightless wind. It stays there because it is being continuously fed by the trees. It is being torn to pieces and scattered by the wind, but at the same time it is being renewed by vapours from below. It is not the same cloud that remains there, but a continuously created one. In the midst of life it is in death, and even as it dies it lives.

It will be noticed that Mr Person is stronger than Mr Zon on the offices of the ocean. For my part I am convinced that there must be a continuous feeding from the ocean. There must be some feeding from the ocean, otherwise it is impossible to see how there could ever be any tree-clouds at all, how there could be anything but drought. I must not pass anything on to the reader which I cannot successfully pass through my own mind. Everyone who writes on these themes is deeply indebted to Mr Paul Sears's *Deserts on the March*; yet he writes (approaching this particular aspect from the other end): 'Forests tend to occur where there is a greater annual fall of rain in inches than the air, on the average, will draw back in evaporation. Where the reverse is true grassland occurs, or if the evaporation is still more intense, scrub and desert.' It beats me how he can suppose that if more goes up than comes down there could possibly be any vegetation at all!

Others will doubtless clear up this misty point as to the ratio of water given, absorbed and returned. In the meanwhile a broad face is clear: namely, that forests by feeding clouds and perhaps making some more on their own, increase rainfall; and that they do this not only for their own locality but for other places since the wind will often carry the vessels a long way before unloading. Thus trees are great distributors of rainfall. Water is not evenly distributed throughout the world and is not always found where it would be most welcome, and trees thus play an important part in its distribution, for not only do they at given places add weight to clouds and bring down the messengers of moisture, but also that vast vapour given off by forests into the atmosphere is often carried great distances, so that trees of one country may be the cause of rainfall in another that needs it more.

We can say this. Supposing that tomorrow there were no vegetation over the face of the earth, then much less rain would fall over the continents; the clouds would frequently pass over without unloading. If on the next day trees covered the same space, then the rainfall would be enormous in comparison. Therefore if continentals (we are not thinking of islanders such as the British, at the moment) wish to be sure of their rainfall, they should be careful about their forests. They have not always been thus careful. The result is that in some places after reckless lumbering, men have looked up to see the clouds steadily passing them by day after day without discharging their moisture, like ships refusing to put into port. The primitives were nearer the truth when they paid special honour and made peculiar sacrifices to certain trees as the producers of rain.[6]

Finally, it is clear from the foregoing that if we are entitled to say, Put up some trees and you pull down some clouds (always supposing the actuality were anything like as simple as that!), we are certainly not entitled to think that this will always add to the moisture of the soil on which they stand. On the contrary, they suck up moisture, as we have just seen. 'The more highly developed is the vegetal cover,' says Mr Zon, 'the faster is moisture extracted from the soil and given off into the air. In this respect the forest is the greatest desiccator of the soil.' As a rule, therefore, the best places for encouraging forests are the hills and the mountains – to mention but one reason. And from this it also follows that trees could be used to suck up swamps and bogs. Swamps are agriculturally useless and often the breeding-places of malaria and swamp-fevers. In fact trees have already been planted for the purpose of draining swamps. It has been done with great success in Landes and Sologne. It would be delightful to see the half-useless peat-bogs of Ireland's Calary Common in County Wicklow transformed in this way.

The above considerations, then, entitle us to say that trees have a decided influence upon temperature; that by offering obstruction to clouds on high places they increase rainfall and in effect raise the height of mountains; that though forests promote a greater fall of rain than do open spaces, they themselves give back almost as much water as they receive, raising invisible

6. See Frazer, *The Golden Bough: The Magic Art*, vol. II, p 46.

oceans which moisten the pastures of the sky and favour a far-flung distribution of rain; and that this very fact enables them to soak up swamps and cleanse their malarial pollutions.

6. The Influence of Trees upon Mountains, Rivers, Floods, Springs, Wind, and Birds

We have not yet exhausted their properties. Trees do more in the economy of Nature. They hold up the mountains. They cushion the rainstorms. They discipline the rivers. They control the floods. They maintain the springs. They break the winds. They foster the birds.

All the barren mountains of the world are falling down. They are too unyielding. They can be overcome and undermined by the soft and the flexible. The wind beats upon them and they do not bend. The rain lashes them and they do not absorb it. The rocks are corroded by the insinuating force of the less substantial elements; the fallen water, finding no entrance in earth, rushes down as on a roof and wears away the flint; the fingers of the wind press ceaselessly against the cracks within the weakening stone. This wearing of the water and the wind, this 'weathering', may be slow; but let no man build a cottage in the valley on that account; for though the gullies take time to carve and the undermining seems insensible, yet tomorrow, today perhaps, a landslide may occur and a big slice of the mountain seek the plain below.

However, most of the mountains are held together by plants (and in some cases by ice). This plant-cover consists of grass or trees or both. Even if the trees do not go up the whole mountain and only stake it down at the middle regions, it still means that a high avalanche will be stayed from total descent by the arms of the trees. 'A mountain without forest is an absurdity,' says Dr Ehrenfried Pfeiffer, 'and it creates a serious illness of the earth's surface.' For trees not only prop up the pile, they attract rain as we have seen. Permit me to quote Pfeiffer in support of what I have just said under that head.

The mountains are the great water gatherers of the earth. On them the clouds discharge their burdens; in the temperate zones and regions of the monsoon, the rainfall increases with increasing altitude, up to a definite point. Curiously enough the altitude approximates to that of the timber line. The precipitation decreases in the region of the Alpine meadows. It is, therefore, more accurate to say that the hills, mountains and forest, combined, are the water gatherers. The clouds strike against the mountains, forests attract the clouds – so speaks the mountaineer and woodsman on the basis of simple daily observations.[7]

Tree-clothed mountains, then, are water gatherers: we have established that. But, again, that is not all of it. They do more than gather the rain. They do more than hold up the mountains. They deal with the rain in the best possible way in the interests of natural economy.

Imagine if you can – it is not something we could ever see in the state of Nature – a long mountain slope consisting of soil without grass or trees on top. When the rain-storm beats down upon it, what happens? Some of the water sinks in and is sponged up; and a great deal more runs down into the valley below to form a torrent making for the lowlands and the sea. The rain-storm continues: and now the water is hardly absorbed at all and ninety per cent of it runs down, *taking with it* a proportion of the top of the soil. The process continues. This very unnatural erosion continues and the rivers increase, while their freightage of soil and silt piles up in the regions far away. The process goes on until all the top of that soil has been carried down. Then the bottom of it follows, the stones, the rubble follow and pile up on the soil which went first, so that matters are now upside down in the valley beyond. The storm subsides. The rivers decrease. The beds dry up. There is a period of drought. Then once again the storm breaks out. This time the water rushes down so unhindered and so swiftly, that what with inadequate river banks and gross siltage, floods sweep over the land.

This never happens, of course, in the Natural Order. It could

7. Ehrenfried Pfeiffer, *The Earth's Face and Human Destiny*.

happen if someone came and rolled up the carpet of grass and pulled out the cover of trees.

Now let us imagine the same area of mountain slope covered with forest. Again the storm breaks out and the rain pours down. This time it does not reach the ground all at once. It must first fall upon the leaves and the branches of the trees, and thence trickle to the bottom where it is easily absorbed. There is no running straight down the hill into the valley. For not only is there so much less force in the rainfall by virtue of the living-leaf obstruction which cushions the blow, but the dead-leaf and twig obstruction, the litter, serves in the nature of a colossal sponge, a single acre of which can sometimes harbour forty-six tons of water. This absorption on the floor is the chief thing and a very great thing, but the check first received at the roof is also important. This is particularly evident in the case of snow, the rapid melting of which is so often a cause of sudden flood where there are no trees. In a dense forest only half the snow-fall reaches the ground. A white roof is formed on top of the trees, so that airmen passing across forest lands have sometimes confused the foliage with the floor. When the false upper floor melts it must first trickle down the barks or fall in lumps to the ground. This capacity of litter to detain moisture is called seepage, which in many places on the steepest slopes has been found so marvellously absorbative that it 'creates conditions with regard to surface run-off such *as obtain in a level country*'.[8] It can turn almost a perpendicular into a flat in terms of gravitation.

It is this seepage which promotes the discipline of rivers, the always wonderful sight of water running on and on all the year round, neither flowing over its banks nor drying up nor becoming clogged with silt. It is seepage which makes severe floods extremely rare in the natural state. It is seepage which preserves the water clean and wholesome for the fishes. It is seepage which keeps the rivers dependable for navigation when men arrive on the scene. Indeed it is already a well-known truth that if we strike at our trees and thus at seepage, we strike at our

8. Paul Sears, *Deserts on the March*.

inland ships; thus (to anticipate the argument for a minute), at the period of Roman rule in France the River Durance was perfectly navigable, while now, the watersheds being cleared of forests, you can hardly float a skiff on it; and the Loire, once a navigable river of the highest order affording communication between Nantes and the Central Provinces, so that in 1551 the Marquis of Northumberland, Ambassador from England, could sail from Orleans to Nantes panoplied in a magnificent suite 'in five large many-coloured boats', is now unnavigable above Saumer, owing to the detritus brought down from the mountains with every flood.

Once, in Jamaica, when in the Blue Mountains, I climbed from a place called The Ark to St Catherine's Peak. Heavy rain happened to fall just as I set out, a tropical deluge, and my path the whole way up became a river-bed. This downpour lasted for exactly one hour and a half. The result was that all the dry river-beds came to life again, and this lasted for about two days and two nights – a steady flow. Though theoretically I understood the properties of seepage I was amazed to see how one hour and a half's deluge could be so sponged up and checked that instead of flooding there was a steady flow fed from within the mountain as from a reservoir. It became easier for me to grasp how an underground system of springs and lakes can be *maintained* in the great forested mountain ranges, continuously fed from the sponges that cling to the rocks. This is the protection against drought no less than the damming against flood.

The consolidation of mountains and the just administration of water do not exhaust the offices of trees in relation to the earth. There is an invisible agent we have to reckon with, necessary and beneficial in the motions of the sphere, but at times a most blasting bane, an unseen foe that needs no cloak of darkness. We must consider the wind on the plain. It is not so drastic an element as water, but it can be a fearful one. Its invisible whips can be the scourge of man and beast and plant. The only thing to do is to break it. An impossible task, we might think. It is easy to break hard things, by simply tapping them or by elaborately blasting them to bits. It is very difficult to break a really soft thing, and when that thing is the unseen element of wind whose arm is yet strong enough to raise up liquid mountains on the sea or cast down houses on the land, the only thing we can do is to wall ourselves away from it. We cannot wall up the open country, so we must try and break that fury. Again we call trees to our rescue, and speak of windbreaks. And it is astonishing to how great an extent they do break it. I have stood on a field protected by a line of poplars when almost a hurricane was blowing across the country, and I have felt hardly enough wind to blow my hat off; while the difference in the temperature between my side and the far side of the trees was remarkable – no wonder, since it is found that even a hedge of only six feet high can raise the average soil temperature three or four degrees to a distance of four hundred and fifty feet.

We take this sort of thing for granted in the British Isles where agriculture is not yet so much the enemy of silviculture that hedges at least are still in abundance. One writes 'not yet' because hedging is a big job in itself, as I can vouch for from personal experience of it, and does not lend itself to any mechanical instrument. But since so many hands have recently been turned into steel, so many men exchanged for machines of one sort or another, the man who looks after hedges may soon be no longer found on a farm; and thus the farmer is increasingly showing a tendency to do without hedges and to put up one long piece of wire instead, charged with electricity for the benefit of the amazed and affrighted cows. I suppose there will have to be a decade of bovine electrical education before these monstrosities are exchanged again in favour of hedges. But elsewhere in the world, on great stretches of plain, neither trees nor hedges are naturally abundant, and 'the wind which sweeps over the plain unhindered, increasing in fury and breadth, is its greatest enemy. In drying out the plain, it creates a hard soil crust. By increasing evaporation, it draws off the soil moisture and cools the soil. Then it tatters and dries the finer soil parts. A plain constantly exposed to wind pressure will be driven back to the most primitive conditions of life and growth.'[9] And should there ever come a time when large areas of level forest are cut and the land ploughed and the soil loosened, then, the hitherto harmless wind in that region will be no longer harmless, and ruined farmers will face that cloud of dust which is their day of judgement.

Thus it is easy to see that the breaking of the wind is nearly as important an office of trees as the distribution of the rain, the ruling of the rivers, and the maintaining of the springs.

One other thing – the birds. We must not forget the birds. They compose one of the forces that serve the rhythms of Nature. Their residence is in the trees. Can we dispense with their services? Without them can we keep down the insects in a natural way? Of course we are always doing it in our own way. We feel bound to do it in the interests of agriculture, especially fruit-culture. If we are to get the benefit of the earth – which belongs to us, we say – we must exterminate a great many creatures who are inclined to eat our stuff. Insects are the worst offenders. Their offence is rank. They must go. Many of them have gone accordingly. The trouble is that either they return by other entrances and perhaps in other forms, or their exit is the cause of increased strength in other quarters. The subject is wonderfully complicated. Probably every single thing in Nature works for the ultimate good of the whole motion of life: if anything, however minute, were an alien performing wrong action, then, either it or the whole universe would surely come to grief. It might be better to keep our trees and let the birds do the job. There are good birds and bad birds from the point of view of agriculture; but is it wise therefore to try to dispense with all of them? The sparrow may be our enemy on the field; but may we on that account dispose of the tom-tit who consumes eighty pounds of agriculturally injurious caterpillars in a summer, or of the cuckoo who eats eight hundred a day? The wood-pigeon may be our curse; but can we dispense with that excellent rodent-eater the owl, or with the starling who is the

9. Ehrenfried Pfeiffer, *The Earth's Face and Human Destiny*.

destroyer of the lepidopterous larvae? These are big questions when we come to weigh them in terms of long policy.

And can we take away the residence of the jay? His specialized knowledge in the best treatment for planting oaks still amazes the experienced forester. With regard to his secret, Dr August Bier has some interesting things to say. When considering the question of how many trees can be left to plant and spread themselves, he reminds us that acorns and beechnuts remain lying under the crowns of their respective parents. And since they do not grow well in the shade of their own species, the forester must operate artificially. 'In untouched Nature,' he continues, 'these trees would have but a limited dissemination were it not for a very ingenious bird who steps in and cares for their spread in a wonderful fashion. This bird is the jay. He carries away the acorns and beechnuts, one in his beak, the rest in his crop, and sticks them into the soil, or far more often into the covering over it, especially into pine needle carpets. And he seems to do this in a much better way than the forester. He reforests evenly over the whole area, never puts several acorns together, but always at correct planting distances, so that a correct and useful stand of trees results.[10] Here and there he also sows in rows, again keeping the correct planting distances ... I wonder ever and again over the fact that the wild pigs let the jay-planted acorns alone while they root up those planted by me, to the very last one, if I do not protect them with fences.' And a little further on he introduces a touch of metaphysics into these deliberations.

In going about the building up of woods according to plan, no creature, outside mankind, approaches the jay; indeed I should like to say that he even surpasses the human forester. And this fact, apparently standing quite alone in all nature, which yet lies in plain sight for every observer, or ought to be in plain sight, is either not grasped in its implications and significance, or is completely overlooked by the experts, even though it is equally important from a practical as from a theoretical point of view. For whoever looks more deeply at once perceives the Logos. When I said that the

10. In Germany the bird is called the Eichelhäher – i.e. acorn-carrier.

example of the jay stands alone in Nature, I quite consciously used the word 'apparently'. In reality something analogous happens frequently in Nature. But again these connections are neither recognized nor attended to by mankind.

For myself, I would prefer to use neither the word 'apparently' nor the word 'frequently'. We should accept these connections as the norm in the Order of Nature.

CHAPTER II

MAN AND THE NATURAL ORDER –
THE OLD WORLD

1. Man Rejects the Order

That we might concentrate our attention upon a few aspects of the Natural Order, followed by a special reference to the activity of trees within that government, as little reference as possible has been made to man, save at the end about wind and birds. It was necessary that we should first contemplate the spectacle of Nature exclusive of men and their activities. What shall we say of that primal state? How shall we value it? It is no garden; yet it is the only Garden of Eden there has ever been. It is no earthly paradise; yet Paradise is there. The earth-refusing idealism of man is at fault, not the creation. 'It is here as in Paradise,' said the old Indian to Alexander Von Humboldt as they went up the Orinoco in a canoe for 1450 miles and frequently watched the many different wild animals coming down to the edge of the river to drink. But Humboldt commented: 'The gentle peace of the primitive golden age does not reign in the paradise of these American animals, they stand apart, watch, and avoid each other.' No doubt; but it is that or nothing. And the solid fact that few of us would choose the void as an alternative to this wonderful world, should be sufficient to prevent us arraigning the justice of what *is*. For here at least we can see that all things work together for good to this extent: that a balance of powers is held in such delicate and amazing equilibrium that water, air, and fire; that plant, animal, and soil live together as if they made a single organ: a balance of performance so policed that no unit ever gets out of hand for long. We are at liberty to call it hell rather than an earthly heaven if we wish; but we should pay careful attention to its composition before we decide to alter it.

This brings us to man. For that is exactly what he decided to do – alter it. He would *not* put up with all that danger, all that discomfort, all that pain and sudden death. He would not be eaten – however much he might eat others. He would not obey the laws of this jungle of Nature, but get outside them. Nature and its primal order was not good enough for him, so he left the Garden of Eden.

He stepped outside. Let us change the pronoun – we stepped outside and became aliens on the organic body of the landscape. What happened from that hour to this is called history. But by history we do not mean the story of man in relation to the earth; by history we do not mean the story of man in relation to the other creatures of the world – that is not supposed to be important. Indeed, we have behaved exactly as if we owned the place; and while carrying out plans for the best use of the property we have provided ourselves with suitable religious and philosophic support by declaring that we happen to be made in the very image of God, and that the proper study of mankind is man. How we have turned to account the whole surface of the earth for our benefit; how we have gone beneath its surface and taken out zinc and copper and steel and gold and iron and coal, transforming them into towns and houses and machines; how we have used any animal that could be employed for our satisfaction and made a mockery of others for our amusement – is called the growth of civilization.

In many ways it has been a splendid story. Many of the forms into which we have cast our effort have been almost sublime. We can never wish that we had not used the peculiar gifts bestowed upon us by nature and had declined to take up the burden of our destiny. Since we rose up out of the earth and were given those powers by the earth, we can hardly be expected to feel overwhelmingly shamed by the irresponsibility of our actions seeing that we were made that way. Yet it is our saving grace that we are capable of self-criticism and even ready to declare that we should be responsible. And when self-interest comes into it we are even capable of mending our ways. We have reached a point now when we must revise our conception of man's place in Nature, or suffer consequences such as had never occurred to us in the process of our conquests. In fact it is not unlikely that in the course of the next few decades – perhaps after we have had some really rude shocks – we will gradually return (this time on the plane of consciousness) into the Order of Nature, as one factor of the whole. If this happens, the turning-point will undoubtedly have been this century.

2. *Agriculture the First Enemy of Trees*

'A landscape,' says Ehrenfried Pfeiffer, 'is an organism, a living entity, possessing organs and functions which react and interact according to definite constant laws.' Its arteries are rivers and streams, its skin mountains and forests and sea. This is a little far-fetched, perhaps; but nearer the truth than the idea of a landscape as a conglomeration of solids and liquids only connected in a haphazard way. In the above section I spoke of man as an alien or unassimilated body upon the landscape of the world. Of course, to say that he left the Garden of Eden to become an alien at once from then on, is not only very allegorical but it is to telescope the history of mankind too severely. He did not begin as an alien body. In fact he began, as we have seen, in so reverencing natural phenomena that he saw deities at every turn, and behaved, especially in relation to trees, almost as if he had scientific knowledge concerning their effect upon water and soil. Nevertheless, the step towards extricating himself from the bondage of natural law had been taken on the first day that he stooped down and picked up a stone with the intention of using it as a *tool*.

Largely, it was all right so long as man remained a hunter. For then his actions with regard to the earth and soil did not differ from the actions of the animals. But after about two hundred centuries of hunting pure and simple, much of it on open steppes, a milder climate and greater vegetation are thought to have stimulated the hunters towards a new idea. I do not know whether it is known exactly where or under what conditions a piece of earth was first *cultivated*. There must have been such a day – the most important in the history of all life on the planet. Now we have so much artifice around us that we tend to forget how artificial agriculture is. A field of wheat is almost as artificial a thing as a piece of pottery; it is organic, of course, but its existence in that form is entirely due to man and would be impossible in the natural state. To make Nature grow things which could then be eaten, thus overcoming the necessity to kill some dangerous animal or starve, must indeed have been an intoxicating discovery. We still see much sense in it. There is no more common sight than a field. And no one, up till now, has ever dreamed of questioning the right or advisability of cultivating a field. Since we owned the earth, we could naturally dig in a spade wherever we wished; that was taken for granted – and it was supposed that a spade could be used anywhere, a plough driven anywhere, with good results for man. 'The cut worm forgives the plough,' said Blake, meaning that Nature is on the side of the agriculturalists. But is she? It is very doubtful whether the cut worm forgives the plough on all occasions.

There is no reason to suppose that Nature is not just as much on the side of trees as of agriculture, and we are not entitled to suppose that the cut worm was particularly gratified when agriculture soon proved to be the enemy not only of all animals who got in the way or who could not be employed, but of the forests. The truth is that trees were the enemies of men in the beginning. Early man was obliged to conquer forests or be conquered by them. When he gave up hunting on the steppes he was trapped in them. 'He could make no headway,' says Mr F. Kingdon Ward.

He possessed neither the tools for making clearings nor the knowledge for raising crops. All the advantages which the tropics conferred on him by reason of their wealth of vegetation and easy climate were wasted. He could be a hunter and nothing more. It was not until very much later, when man was already civilized, that he learnt to subdue the tropical vegetation. One has only to compare the culture of the North American Indians with that of the tribes inhabiting the Amazon basin to see what this implies ... For the jungle conquered primitive man. Centuries were to pass before man conquered the jungle – if he *has yet conquered it*.[1]

It was not impossible for man to be assimilated organically into the landscape so long as the population was low. It was extremely low in the hunting days. A family then required and received a living-space of fifty square miles. Owing to cultivation the Nile valley was eventually made to support one thousand persons to the square mile; and in China they sometimes managed to bring the figure up to seven thousand. Ah, it is a dread subject, this of population! I confess that when I think of it I am in danger of stylistic inelegance. It seems to vitiate all one's hopes – nay, all the good works of good men, especially of agriculturalists. It promotes such dreadful nonsense and such dreadful cant. The exhortation to 'increase and multiply' was pronounced to a race consisting of eight persons – not to the Chinese, or the Russians, or the Indians, or the Americans, or the British. Our approach is so insolently illegitimate. We talk as if we had a right to have huge populations – whether in terms of a nation or of mankind generally. We have no more right to do this than any other species. Not that Nature cares. Whenever any species oversteps the mark in numbers or behaviour, it pays the penalty and meets with catastrophe. We consider this quite in order and recognize it to be excellent when some huge natural slaughter by starvation or cold overcomes some obnoxious tribe of insects. We approve the scheme. In 1770 the vastly over-populated continent of India was the victim of a famine in which ten million people died. That was excellent – as seen from the viewpoint of the animals. It was secretly thought to be excellent by other nations. Our turn may come tomorrow – and it will be seen as excellent by others. Nature keeps this book balanced all right, since she has all eternity to work in. But our approach is so extraordinary. We really do seem to think that human beings should be exempt from

1. F. Kingdon Ward, *About This Earth*.

natural laws. We even speak of 'the sanctity of human life'. We never dream of extending this amiable ideal towards our fellow creatures – and in any case the cant phrase is cast aside the moment we start a war (though modern wars have been much more destructive of property than of persons). We are perfectly ready, by means of our colossal numbers, to despoil the whole earth and use or mock or kill every other living thing thereon while not expecting that as a consequence we should suffer at all.

It is many years ago since Havelock Ellis wrote:

There are people among us, and not a few, who view with complacency the vast increase of the world's population everywhere taking place, people who would even urge the human procreative impulse to still wider excesses. Until every square yard of the earth is intensively cultivated by Man, until the virulent air is soaked with the noxious fumes of human machinery, until the sea is poisonous with human effluvia, until all earth's shores are piled high with the sordid refuse of human maleficence, it seems to these people that the world will never be happy.[2]

The voices of such people are not so often heard today, though in England (whose vitality is severely sapped by over-population) there are always those who advocate still more births in the name of Power. We must have more people of killable age to fight for us in case of attack, they say. I see the point. But it is like saying: Let us destroy England in order to save her.

Apart from this provincialism, the danger and the horror of an over-populated world are beginning to be realized. But does the realization make the slightest difference to the rate of increase? The rough statistics are startling: according to Sir John Boyd Orr there are 300 million extra mouths to be fed every three years. It hardly bears thinking about if we cast our minds forward a mere ten years. Supposing we put our entire agricultural and silvicultural situation on to a perfect ecological basis, and produced far more of everything than we do now – just supposing – what good would it do if the crop of human beings

2. Havelock Ellis, *Impressions and Comments*.

increases at the same time? It staggers me when I hear eminent agriculturalists like Sir John Russell and Mr Ralph Wightman saying cheerfully in public that things are not so black since we can easily produce more. No doubt we can. But what good will that do if an expanding agriculture only means an expanding population? Then all their plans, all their work, all their knowledge go for nothing. That is why I say this is a dread subject – it vitiates our hopes and takes the vigour out of action. Still, let us not fall into despair over it. I admit that it sounds worse in the aggregate. If *each nation* faces its own population problem, fearlessly and uncantingly, it can solve it on a sound ecological basis; and if any nation is lunatic enough not to do this, that is its own funeral – and it will certainly have funerals.

To return. In early days things worked out all right, even after hunting had given way to agriculture, so long as men remained ignorant and superstitious. They were humble and afraid of offending the deities resident on earth. They were free from Freedom from Fear, which means liberty to exploit or destroy anything you do not reverence (and nowadays there is only one natural object reverenced by man). It meant that they were obedient to natural laws. Today this is called ignorance. The thing is to use natural laws for our own ends. This is called 'ameliorative improvements'. This is where a little knowledge is a really dangerous thing. When man started by living in the setting of a simple huntsman he knew little or nothing of Nature's laws, yet could not help conforming to them as did the plants and animals around him. 'Gradually, however,' says Mr Paul Sears,

and with many halting steps, man has learned enough about the immutable laws of cause and effect so that with tools, domestic animals and crops he can speed up the processes of Nature tremendously along certain lines. The rich Nile valley can be made to support, not one but one thousand people per square mile, as it does today. Cultures develop, cities and commerce flourish, hunger and fear dwindle as progress and conquest of Nature expand. Unhappily, Nature is not so easily thwarted. The old problems of population pressure and tribal warfare appear in newer and more horrible guise, with whole nations trained for slaughter. And back of it all lies the fact that man has upset the balance under which wind and water were beneficial agents of construction, to release them as twin demons which carve the soil from beneath his feet, to hasten the decay and burial of his handiwork.[3]

3. The Reply of the Trees:
The Buried Cities in the Desert

There has been no monumental study compiled, linking the history of mankind and his civilizations with the reactions of Nature. No history of ecology, in fact. It is not until modern times that we have become aware that such history is the most important of all, and perhaps the kind of history that we can learn from. But previous to the modern era we are rather in the dark concerning man's successes and failures to set up his rule within the government of Nature. When today we are considering the future fall or stability of a civilization, our thoughts go straight to the respective soil situation. With regard to the past we cannot be dogmatic; but it is clear that long before machines arrived we fell foul of Nature in many places on a large scale.

Nevertheless we should go carefully here. We must not suppose that every scar or waste place upon the earth's surface is due to erosion caused by human beings. That is going much too far. I fear that too many of our soil erosion prophets of woe – especially in England – are fond of making sweeping statements of this kind unsupported by the slenderest proof. There seems to be no limit to the rhetoric of their meditations. Gazing upon the watered waste of the Atlantic Ocean now flowing upon what once had been an arm of land, they are 'unable to resist the belief' – that is, they wish to believe – 'that the prime cause of the Atlantic disaster was deforestation and erosion.' Men who can think of trees only, cannot see the wood for the trees. They will stop at nothing. They remind me of the critics of

3. Paul Sears, *Deserts on the March*,

Shakespeare in general and of *Hamlet* in particular: having adopted a standpoint they will say anything in support of it. Having claimed the sinking of the Atlantis for the deforesters, they then point to the grandness of the Grand Canyon of Colorado as further evidence of Nature's reply to the transgressions of men. They flatter us, I think. We can hardly claim the canyon as our doing – we must hand it to God. When we stand upon a floor in Heaven, as it seems, and look over the edge down into those stony gulfs of earth, we know instinctively that nothing done or left undone by man has caused the abominable beauty of that rugged way. Anything set going by man must be thought of in terms of time, whereas here, if the agent is not earthquake, then we look upon a playground of eternity for the motions of the air and the wearing of the water.

Again, we must be careful when we speak about deserts. It is not true to say that all, or anything like all the deserts have been caused by man. There are some fifty-six million square miles of land, of which about twenty-two million square miles is desert or near-desert – that is, two-fifths. The root cause of deserts, of course, is drought; and drought depends upon rainfall; and rainfall depends upon the direction of air currents. If the right air currents do not pass over a given tract of land, then neither man nor tree can prevent drought. In such places neither reference nor skill can avail those who attempt to live there; they are at the mercy of the merciless and withering scourge of drought. Man is not the cause of all the sealess seas of sand; he is not responsible for every field of desolation where neither man nor beast nor plant can find a footing. The full story of the deserts is unknown. Tablets of the tale can scarce be furnished by the yielding sand. No fossiled script can be unfolded. Here are no books of stone. And yet at any time the amazed adventurer may come upon a loose page scriptured by a lost city or a forest turned to marble. 'I stooped to waken a sleeping Bedouin and he turned into the trunk of a black tree,' writes Monsieur Antoine de Saint-Exupéry whose aeroplane had crashed in the desert between Cairo and Alexandria.

A tree-trunk? Here in the desert? I was amazed and bent over to lift a broken bough. It was solid marble. Straightening up, I saw more black marble. An antediluvian forest littered the ground with its broken tree-tops. How many thousand years ago, under what hurricane at the time of Genesis, had this cathedral of wood crumbled in this spot? Countless centuries had rolled these fragments of giant pillars at my feet, polished them like steel, petrified and vitrified them and imbued them with the colour of jet.[4]

The history of this forest which once had bloomed in that now desolate place might well have to speak of its growth, its empire, and its blasting, long before the arrival of man. But it is true also that we can stumble upon petrified cities buried in the desert. It is a matter of cold fact as stated by geographers that 'the remains of ancient cultures, buried cities, and abandoned sites, going back at least five thousand years, have been found in most of the old world deserts'.[5] Consider the meaning of those words. They do mean, they do say that here! and here! – is now plain sand where once the waving grass was green; here! and here! – a lawn of dust with but a mast to mark the ship-wreck of a city. We are impressed by the completeness of such shipwrecks when we think of the lost city of Cuicul in North Africa, whose temples and churches, forums and storage pavilions were only discovered by the sign of a solitary column standing erect three feet above the surrounding silt; or of the perished Thydrus in Tunisia which once held a coliseum to seat sixty thousand spectators; or of Thamugadi at Timgad, founded by Trajan in the first century, whose carved porticoes, public library, Roman baths, imperial theatre, and marble latrines lay lost under the waves of sand for one thousand two hundred years until a portion of an archway and the crown of three columns caught the eye of a traveller. What happened? We must beware of putting everything down to deforestation or crop overloading. In some cases the climate may have changed and the air currents failed in their offices so that the clouds

4. Antoine de Saint-Exupéry, *Wind, Sand and Stars*.
5. F. Kingdon Ward, *About This Earth*.

passed by without unloading those drops which give sentence of life and lacking which the cities are as trash and trinkets – though some authorities declare that 'there are no real evidences of geological desiccation of a region, namely from natural causes, within historic times.'[6] But at this distance of time we cannot pass judgement in all cases upon the cause of these conquests *by* nature.

Though we may not have created all the deserts, it is unquestionable that we have created some of them, and added to others. It is not certain whether the once fertile regions of portions of the Sahara were desiccated by the folly of man or the finger of Nature; nor can we say for certain whether the once flowering section in the Gobi near Turfan, regarded as the cradle of civilizations, was turned into dust by man-made or natural erosion. We do know that four thousand years ago Iraq was the granary of the ancient world when the River Tigris enriched the empires of Babylon, Assyria, and Nineveh, as Ur, Khidabu, and Nippur were fed by the Euphrates; and we know that today 'the Tigris flows menacingly on a raised bed of eroded soil brought down from the hills when the plainsmen, seeking more water for their irrigation crops and more land to replace their exhausted soils, cut down the hill forests and were rewarded with uncontrollable floods that overwhelmed their fields and swept away their irrigation works.'[7] That land of Mesopotamia is said to have been one of the most productive in all history; it could support those Babylonian and Assyrian splendours which have become so fabled that they are veritable cities of imagination enthroned in the memory of mankind. But the time came when no captives could hang their harps upon the willows of Euphrates' stream and sit down and weep by the waters of Babylon – for there were no waters: the cities fell, and for the same reason as fell from glory and plenty the Holy Land, the land that flowed with milk and honey; and it was not until the great Allenby, of massive sympathies and searching intellect, stooped down to plant trees even in the

6. Fairfield Osborn, *Our Plundered Planet*.
7. Jacks and Whyte, *Vanishing Lands*.

middle of his campaign, that the fatal deforestation of Palestine began to receive attention and the waste lands to flower and shine again. We do know that before the fall of the Roman Empire a girdle of forest reached in Africa from the Congo to Khartoum – now changed by deforestation and resulting silt to an area of desert covering one thousand two hundred and fifty miles. We do know that soil exhaustion, crop failure, and land abandonment encouraged a steady encroachment of the desert on the fringes of the Persian Empire until sand and mud conquered the conquests of the mighty Darius. We do know that the same desolations of dust now rule in place of the weakened earth that had been so rich around Carthage in the days of Hannibal, and cover the very spot where, lifting the poison to his lips, the hounded leader said: 'Since they cannot wait for the death of one old man . . .'

All this was early days. We shall see presently how in modern times, Man, so much more speedy in spoliation and skilful in ruin, can lay waste the surface of the earth.

4. The Reply of the Trees:
The Lost Cities in the Jungle

The lost cities and lost civilizations of the world have not always been found beneath the deserts. Sometimes it seems as if the jungle had advanced or returned upon them, and like an armed host, bid them quit. Thus with 'the glory that was Maya'. In 1519 Hernan Cortez, the great Spanish conquistador, stormed and took the city of Mexico-Tenochtitlan. He did not know then, and he never knew, that it was built upon the ruins of the Mayan civilization. Five years later he travelled across what is now the little republic of Honduras, hacking his way foot by foot through an almost impenetrable forest given over to reptiles and insects and the odours of putrefaction. Had he turned aside from the path he was cutting, by only a fraction, he would have come to a little stream where he would have found in the midst of all this luxuriant foliage the ruins of what had once been a great city.

It was the City of Copan, the chief light amongst others such as Tikal or Palenque of the Mayan civilization which existed between A.D. 176 and A.D. 620. They are still, far from all other human habitations, lost in the powerful tropical forest which 'like some sylvan boa-constrictor, has literally swallowed them up and is now devouring them at its leisure, prising the fine-hewn close-laid stones apart with its writhing roots and tendrils.'[8] Copan had been seven miles long by two miles wide, with streets and courts paved with stone, a system of drainage and sewerage and with high central temples, palaces and towers. Not only was the architecture evidently most massive and imposing, but the sculptured decorations pass through an obvious evolution from crude and angular carvings to a sculpture marked by purity of style and straightforwardness of presentation, winding up with a period of flamboyance. And the people who had built this city and others like it, had also devised a system of writing by means of hieroglyphs, and an intricate calendar based upon elaborate astronomical calculations.

These cities were built to last, they were not made to be abandoned. But they were abandoned. By the seventh century those streets were silent, those courts were still. The jungle returned. Nature took back her fields. Like besieging soldiers the creepers climbed the walls. Groves grew upon the roofs of the pavilions. Wild vines trellised the rafters of the haunted halls. The massive pyramids sank into the greenery below. Not just one city but all. The glory that was Maya went down into the growing gloom.

Where did the people go? Evidently they trekked north into Yucatan, and established there an inglorious Second Empire until the coming of the Spaniards. And why did they go? They have not told us. Their descendants could not even read the hieroglyphs. This ignorance is to us an amazing thing. Nothing so much differentiates us from early ages as this matter of

8. A. J. Toynbee, *A Study of History.*

consciousness of what is happening and what has happened. Today anyone can know anything that is happening anywhere. Our sense of history, present and past, is like a great network of awareness drawn over us and knit by endless words. In former days a whole people could abandon their cities and move to another land without leaving any record of the cause – not even one tear-stained letter of a child who had lost her doll in the process. Nor any questions asked, answered, or noted down by other races of the day, who might have been interested in so enormous a happening involving what must have been great suffering and grief and shame.

And still we ask, why did they go? It could not have been due to pestilence – for they would have returned. The solutions advanced today are all ecological – they spoilt their environment and could not feed themselves. It does not matter how well organized a society may be, how massive its buildings or good its art, if there is not enough to eat. It is known that when they came they possessed very sharp axes with which they were able to cut down the forest. It is known that having cut them down they then burned them and planted crops in the ashes. So for some time there was abundance of maize, cacao, beans, and other plants which did very well in that splendidly fertilized soil. But the organic material, now open to excessive heat, evaporated much of its vitality in gases and lost much in the soaking wash of the heavy tropical rains. Fresh clearings had to be made and the old ones abandoned, so that step by step cultivation advanced towards ever further fields. Such is the modern ecological solution. 'The Cities of the Mayas,' says Paul Sears, 'were doomed by the very system which gave them birth. Man's conquest of nature was an illusion, however brilliant. Like China before the Manchu invaders, or Russia in the face of Napoleon, the jungle seemed to yield and recede before the Mayas, only to turn with deadly, relentless deliberation and strangle them.' Mr Aldous Huxley takes much the same view in *Beyond the Mexique Bay* when he writes: 'The clearing of the forest led to erosion, and in course of time all the soil was washed from the fields into the lakes. The result was doubly disastrous: the fields became barren and the lakes turned into enormous mudholes.'

If we are to understand the calamity we must use our imagination and extend our sympathy. When they came they were faced with a formidable foe, as they saw it – the thick tropical forest. To have made any clearing at all would have been an achievement. They not only did so but built those cities; and we must see in our mind's eye, first the forest as it is again today, and then see it transformed into great buildings in the midst of populous towns surrounded by cultivated fields. What will and courage and force! It is not strange that it never occurred to them that the foes which they struck down would soon rise up again and advance upon them and drive them back. Yet so it was. The very methods which brought them victory and gave them increase led to their undoing. 'The transitoriness of human achievement and the vanity of human wishes,' says Toynbee in this connection, 'are poignantly exposed by the return of the forest, engulfing first the fields and then the houses and finally the palaces and temples themselves.'⁹

Thus when Hernan Cortez passes along their way nine centuries later, he does not observe the great city of Copan hidden in the knotted mash of the foliage through which he is slowly hacking his path. Such drama can scarcely enter the theatre of our thoughts. It is too big. We turn away. We can stage Hamlet and weep for him – even for Hecuba. But not for the Mayas. Their play is too dramatic. We cannot really believe it. And yet we must. For we belong to a generation which having seen the conquest of nature reach unexampled triumphs now looks round with alarm at the possibility that the pillaged earth may not be able to support us much longer. And in these days, accustomed as we are to so many bombed sites in cities, the idea of Nature covering up the works of man is easily appreciated. Nothing has seemed more striking to Londoners than the extent to which, in a matter of four years, the visiting plants have trellised with their living green the formless husks of ruin.

9. *A Study of History.*

5. The Reply of the Trees:
Floods and the 'River of Sorrow'

This idea that we shall have to mend our ways in dealing with the earth if we mean to remain on it ourselves is something new. We look back across the great civilizations of the past and find that even the greatest of them fell short in ecological understanding. This is true even of China. Of course, the peasantry of China has commanded the respect of all historians and all travellers. 'There is no other peasantry in the world which gives such an impression of absolute genuineness and of belonging so much to the soil,' wrote Count Keyserling in his *Travel Diary of a Philosopher*. 'Here the whole of life and the whole of death takes place on the inherited ground. Man belongs to the soil, not the soil to man; it will never let its children go. However much they may increase in number, they remain upon it, wringing from Nature her scanty gifts by ever more assiduous labour; and when they are dead they return in child-like confidence to what is to them the real womb of their mother. And there they continue to live for evermore.' And again: 'The land represents at the same time one great cemetery of immeasureable vastness. There is hardly a plot of ground which does not carry numerous grave mounds; again and again the plough must piously wind its way through the tombstones.' And once more he says: 'Every inch of soil is in cultivation, carefully manured, well and professionally tilled, right up to the highest tops of the hills, which like the Pyramids of Eygpt, slope down in artificial terraces.'

Much as I admire the Count, I cannot think that what he says here has in it as much soundness as sentiment. It is true that the Chinese maintained the fertility of their land for four thousand years. It is true that they manured with a careful eye to the law of return, their animals giving eight tons each a year, and every adult contributing excreta (night soil) to the measurement of an annual thousand pounds. Even so this is not everything. To live on this earth comfortably requires more good sense than they showed. Ecology is the art of living happily with Nature. They did not display this art. Far from it. They produced so many *of themselves* that they were always compelled to live at starvation level. Their numbers varied of course, but the population sometimes reached six thousand to seven thousand per square mile! That is to say they sowed an extra crop, not of rice or sugar cane, of bananas or citrus fruits, but of human beings. They did not harvest this crop, and then sell it, or eat it. They allowed much of it to rot, and then ploughed it in. For periods of drought could not be faced by so many. Between 108 B.C. and A.D. 1911 there were one thousand eight hundred and twenty-eight famines, in the course of which several millions perished. In the famine of 1920–21 the death-roll was five hundred thousand, while twenty million were reduced to such hunger that they ate flower seeds, poplar buds, sawdust, thistles, leaf dust, elm bark, roots, and stones ground into an artificial flour as an aid to the digestion of withered leaves. This is scarcely the art of living.

Nor is this all. We cannot possibly say that they did well by nature. They did very ill by their trees. The deforestation of the Chinese uplands provides one of the worst examples of ecological stupidity. It is historically and geologically certain that what is now China proper was very heavily forested, and that nine-tenths of the Shansi mountainous area was tree-clothed. The trees were cut down. If there is one thing more famous than the marvellous Chinese agricultural terraces which did so much to control erosion, it is that erosion itself of the rich loess in the north which, when exhausted and loosened by removal of wind-breaks and root-grips, swiftly and massively eroded, carving gigantic furrows on the face of that fertile earth until at last it came to look like a blasted battlefield 'scarred by forces far more destructive than any modern engines of war.'[10] It is not surprising to us – so wise after the event – that subsequent terrible droughts were interspersed by periods of uncontrollable floods. The loss caused by the undisciplined Hwai River, which in 1911 flooded an area larger than the size of Belgium, is said

10. Jacks and Whyte, *Vanishing Lands*.

to be sufficient to have supplied food for ten million people. They have been obliged to hold in the rivers by means of earthen dikes; but on account of the colossal amount of silt brought down, these dikes have had to be continually raised so that now the bed of the river is above the land and liable to spill over – as witnessed by the floods in Chihli, Shantung, Honan, Kiangsu, Anhwi and Chekiang. As for the Yellow River or Hwang Ho whose very name means 'China's Sorrow', it is now as great a scourge as the Yellow Peril. It may be a glad colour, but it has a sad cause. It is China flowing away. Every year two and a half billion tons of China's soil is carried into the sea by this service. This means twelve inches yearly taken away from an area covering one thousand two hundred square miles. Thus is Nature avenged. Thus have the trees replied. An armed host of men came out against them and laid them low; and even as they fell, their perishing roots that had gripped the earth and checked the waters, were loosed, and the sins of the slayers were visited upon many generations of their children.

It is not to be denied that China built up a wonderful civilization upon the backs of the peasants. It is a big continent, and it was possible to go on for four thousand years before Nature really showed her hand too severely. A mighty Wall could be built and a marvellous feeling of security prevail in the cities whose elaborate palaces with their Pavilions of Feminine Tranquillity or of Charity Made Manifest provided a suitable setting in which Empress Dowagers could wear capes decorated with three thousand five hundred pearls, and the Son of Heaven order a bowl of soup costing ten thousand pounds. The mandarins, the eunuchs, and the literati could subserve their masters and issue their elaborate memorials in a style worthy of a culture that should last for ever, while the pottery, the poetry, and the misty miracle of their pictures were perfectly congruous with the moral maxims of their sages. Master Kung might meet with opposition but not with indifference, and through the nobility of his demeanour and by the pearls of great price which he let fall from his lips as he wandered on his pilgrimage, he could establish an ideal of behaviour which seemed possible of fulfilment in so stable a civilization. The lofty exhortation of Gautama in the annunciation of the Law of Right Effort: 'Strive to avert the spreading of evil that hath arisen: strive to avert the arising of evil that hath not arisen: strive to aid the spreading of good that hath arisen: strive to aid the arising of good that hath not arisen': could fall into place in the framework of that day. No civilization in the world has ever been more fitted to respond to the quiet wisdom in the question and answer of their great philosopher, Lao-tse: 'Who is there who can make muddy water clear? If you *leave it alone* it will become clear of itself.'

Total political revolution became inevitable in China. One great advantage of their new scheme of government is that when a given line of action is decided upon it can be put through, and neither opposition nor lethargy are permitted to undermine the policy. It remains to be seen to what extent the Chinese government is now ecologically determined.

It is right in these perilous modern days when the top of the earth is nearly everywhere showing signs of exhaustion, we should turn an ecological eye upon the fall of empires and the mystery of silenced cities buried in the desert or lost in the jungle. Perhaps we have had enough purely historical speculation. And there has been enough denunciation on strictly moral and religious grounds. The wickedness and luxury of cities, the pride of rulers, the infidelity of an ungodly and sinful people, have always made a favourite theme for angry prophets and holy men. Saint Paul preached in the City of Antioch against its pride and its sins. There were four hundred thousand people then in that city whose Pleasure Gardens of Daphne were the envy of the Mediterranean world. Today it is merely a drab little town in Syria whose former grandeur has been reconstructed after digging through eighteen foot of detritus caused by the silt in the Taurus and Lebanon Rivers carried down when the protective terraces had been neglected and the spoliation of the forests completed. 'There is after all,' says Mr Stuart Chase, 'no philosophical difference between the fate of Antioch

in Syria and the possible fate of the Garden City of Kansas.'[11]
No doubt Saint Paul was right to preach against the people of
Antioch, and other prophets to lay their curse upon other cities.
But they did the right thing for the wrong reasons. Those sins
were not moral; they were not theological — they were ecological. That pride and that luxury might have been a great
deal more pronounced and yet no harm befallen them; the
green fields would have continued to yield them increase and
the limpid water to bring refreshment; that immorality and that
impiety might have spread further and mounted higher, and
still the strong towers would not have shaken and the massive
walls would not have crumbled; but because they had been
unfaithful to the land upon which they lived; because they had
sinned against the laws of earth, and despoiled the forests, and
loosed the floods, they were not forgiven, and all their works
were swallowed in the sand.

11. Stuart Chase, *Rich Land, Poor Land.*

CHAPTER III

MAN AND THE NATURAL ORDER – THE NEW WORLD

1. The Extermination of the Indians

Three hundred years ago the whole of the North American continent was six thousand years behind European civilization. It was only inhabited by Red Indians, and not more than a million of them, while long stretches of wild meadow and primeval forest, extending like years into the distance, had no human dwellers at all. A wildly beautiful land, enormously fertile, carrying but a million Indians – it is difficult to conceive now. When we remember the Red Indians and the things they did not want, the clothes they did not wear, the houses they did not build, the roads they did not need, the laws they did not make, the goods they did not sell; when we consider how they could see in the dark, how they could run swifter than wild horses, how they could wrestle with the eagle on equal terms, how they could hear over immense distances, how they could run naked in the snow and frost without feeling cold; how they lived with nature from sunrise to sunset; when we consider these things and then think of the modern civilizations in general and of America in particular, does it not seem that machines have taught us to pursue one goal above all others – *comfort*? Part of natural life is discomfort. Can we really eliminate it?

In the compass of three hundred years that wonderful continent was Europeanized – only rather more so – and plastered with towns and tied up with roads and machines, while the population increased by over one hundred and fifty million. And now, today, after that three hundred years, we hear voices in extravagant prophecy, declaring that in *one hundred years* there will be no soil left on the top land, and that it will not support a single man.

But not so fast. Let us proceed quietly.

We have just passed in review some of man's failures to live within the balance of nature. Perhaps the most obvious example of a culture *not* working at her expense and therefore not suffering shocks, is that of the North American Indians. 'Nature needs to be let alone,' writes Mr Donald Culross Peattie, the American naturalist and gifted writer. 'Free to her own devices, she cleans her own house, knows no wastage, makes no biological mistake. She solves for herself the problem which men are still finding insoluble – that of a balanced economy.' And the Red Indians let her alone.

The red man never dammed a stream, never drained a swamp, never exterminated an animal. What ground he cleared for his primitive agriculture was negligible. On the prairies he lit fires sometimes to round up the game, but the only lasting result was to keep the Appalachian hardwood forest wall pushed back to the east, preventing it from encroaching upon the prairies, the great meadow, the American steppe-land on which the bison herds depended for their lives. In no way did the Indian break the charmed circle of the wildlife community.

It is not my task here to appraise or dispraise their remarkable social life and religious sense which open up vistas not only of beauty but of horror and cruelty. Our concern is with the Indian attitude to Nature. It was the polar opposite to those who came from Europe to destroy them. They fed themselves very largely by hunting – becoming quadrupeds themselves in pursuit of the bison, by dressing in wolves' clothing. Their agriculture was largely confined to maize, which, so far from being regarded merely as 'a crop' was the great Mother. Just as the sun was not merely 'a ball of incandescent fire' but (for some of them) Uiracocha the Lord of Reproduction, so an ear of white maize, with its tip painted blue to represent the sky, and with four blue lines running down it, symbolized the dwelling-place of the Spirits and the four paths by which they would descend to minister unto man in answer to the prayer uttered at the Procession of Peace: 'Mother Corn, Oh hear! . . .' And the soil was not less maternal than the maize. When Smohalla of Nevada received agricultural instructions from Washington he replied: 'You ask me to plough the ground. Shall I take a knife and tear my mother's bosom? Then when I die she will not take me to her bosom to rest. You ask me to dig for stones! Shall I dig under her skin for bones? Then when I die I cannot enter her body to be born again. You ask me to cut grass and make hay and sell it and be rich like white men. But how dare I cut my mother's hair?' Nor was the veneration of trees by the Indians any less pronounced. Believing that every natural object possessed its spirit, they attached special importance to the shades of trees and approached the cottonwoods as if they were beings of higher intelligence than themselves, and believed that misfortune would follow if due respect was not paid to them, while their respect for the acacia was so great that the offerings of blankets and ribbons and ponchos, and the gift of tattered garments drooping from the boughs often made it look like an old-clothes shop.

It was shortly after the above reply by Smohalla that the Hotchkiss guns of the invaders were trained upon the camp of the Sioux tribe, pouring two-pound explosive shells at the rate of fifty a minute, mowing down everything alive, until two hundred Indian men, women and children were lying dead or wounded on the ground while the remainder fled into a ravine to be pursued and massacred.

Mr Harold Nicolson tells how when lecturing in America he was generally asked by someone in the hall after a lecture: 'What about India? How can you justify the British behaviour to the Indians?' At last exasperated one day, he turned and replied: 'What Indians? Ours or yours? We educated ours. You massacred yours.' After a painful silence of astonishment at this rejoinder, Mr Nicolson says, 'they had the decency to laugh.' An excellent exchange. And yet one cannot help feeling that the proceedings would have been rounded off more completely if someone from the body of the hall had then asked Mr Nicholson: 'Who do you mean by you?' For it happens that there were no Americans in those days, only Europeans – and chiefly Anglo-Saxons who, stimulated by the initial example of the

Pilgrim Fathers who first falling upon their knees, as the saying goes, then fell upon the Indians. We are about to consider the way of Americans with trees and soil, but we cannot say at what exact point we are entitled to dissociate ourselves from them and draw back the hem of our garments. But of course the Americans are proud of the early achievement in conquering the wilderness and do not wish to be excluded from the aspect. A good way out, adopted by Mr Peattie, is to refer to 'our ancestors.' Thus he says:

The task was to clear a space around the coastal settlements; to leave the Indians no lurking place; to push back the toppling green wave of the forest; to give the dreamer's mind room to think. Then the pioneers planted the seeds of civilization. Our ancestors, unlike the Spanish in the New World, did not bring civilization to the natives: we got rid of the natives and kept civilization for ourselves.

When they had got rid of the natives it only remained to take over the property: A new, vast, uninhabited country lay open before them. Endless forests and long-drawn virgin vales such as had never before and would not again confront the eye of an invader, were theirs for the taking. 'They entered an Eden such as the world will never see again, the last unspoiled wilderness of the temperate zone, teeming and complete with a life of its own.' The empire of the trees occupied an area larger than any European country save Russia. Rooted in deep, rich primeval loams, the beech forests of the Middle West produced enough nuts – a billion bushels it is said – to feed the passenger pigeons who flying a mile a minute hour after hour sometimes for days on end, clouded the sky by their prodigious numbers.

The Indians – few in any case – dwelling within the gigantic sweep of the plains had formed no concentrated settlements. They had possessed no machines. They had used no ploughs. They had kept no cattle. Food was for use and not for sale. Thus there had been no need to make demands upon the soil, and they had always preserved a sense of their dependence upon the natural environment. So when the whites achieved vacant possession they found one of the most bountifully endowed continents in the history of the world, in perfect condition. Nothing had been lost in the process of natural erosion – neither soil itself, nor minerals in the soil. The 820 million acres of forest held the continent in a grip which neither water nor wind could possibly disturb. The 600 million acres of grassland were no easier to shift. Nor was there any danger of losing the 430 million acres of open woodland. The desertscape, which amounted to 50 million acres, was held at a minimum by the play of forces in the natural equilibrium. The rainfall, whether gentle or torrential, was forced to pass through a scheme of drainage which ensured that its distribution was expansive, its flow orderly, and its composition beautifully pure.

Perhaps the continent had remained like this for some twenty thousand years. Then in 1630, that famous ship, described by Milton as built in the eclipse and rigged with curses dark, landed upon the shore. The passengers came with big ideas. They saw at once that this Order of Nature was not good enough. They must exchange it for one of their own – with their convenience as the criterion of its excellence. In three hundred years they achieved this.

2. The Campaign Against the Forests

Having conquered the Indians, they turned to Nature. They found themselves confronted with a mighty host. It stood before them, erect and menacing, battalion behind battalion. But it was unarmed. It could not defend itself. It could not even retreat, for it was rooted to the ground. Being pious folk, the invaders saw that God was clearly not on the side of these green battalions. The forest was an enemy that could be destroyed. And they set to work to destroy it.

Two immediate objectives were to be gained: first, room in which to grow crops, and second, the supply of timber with which to build the new civilization and to maintain it with fuel. It was a big job, this subduing of the wilderness. It took toughness and time before the first 100 million acres of trees had been brought down. They went at it with a will. They launched a campaign against the forest with a virulence that seemed akin to hatred (an attitude towards trees which seems to have remained with Americans to the present time, for in 1917, according to Dr Pfeiffer, when some French peasants asked American soldiers to thin out a few trees, they were appalled by the recklessness and violence with which they set to work). They went out against the forests with the thoroughness of an invading army, attacking first one stronghold then another. For a hundred years the white pine trees of New England held out. Then one day it was found that all had fallen on that field. After the white then the yellow. The movement of destruction advanced relentlessly onwards from the forest of Maine to New York. In ten years those battalions were defeated and the lumber-troops entered Pennsylvania and Ohio and Indiana, from whence they moved in turn to Michigan, to Wisconsin, to Minnesota, and thence again through the Rocky Mountain region on to the Pacific Coast. That was the northern campaign against the trees. There was a similar offensive in the South, from the Carolinas to Texas and on through Arizona and Colorado.

Such is the briefest possible outline of the onslaught. Today it is reported that seven-eighths of the continent's virgin forests has gone, and that only the Douglas fir is making a last stand along the fifty-mile front between the top of the Cascade Mountains and the Pacific.

The method throughout this tree-war was that of clean cutting, complete clearing without any policy of further yields. It was applied to areas which had no farming possibilities with as much zest as on the fertile loams. Whole mountain ranges were burned off, though quite useless for farming. The attackers advanced upon the enemy with steel and fire – with quite as much fire as steel. Burning down forests by deliberate intent was one of the quickest means of advance, for in a wind fires sometimes spread at sixty miles an hour. Speaking of the Southerners, Paul Sears says:

To the settler, here as in the North, the forest was a hostile thing, occupying the ground which he needed for corn and beans, even though it furnished him with game, fuel and building material. All was fair in the struggle against this handicap, and no weapon, not even his sharp axe, was more powerful than fire. So the use of fire against the forest became a ritual of the poor white. He has literally burned his way west, from the pine-lands of the Carolinas to the black-jack cross-timbers of Oklahoma and Texas.

There seems to have been no sense of waste in those days. Often enough they did not stack the logs for later use – it was easier to get them out of the way by burning them at once. Thus it is told how, in the neighbourhood of Michigan, huge slabs of white pine were dumped into the open fields in great pyres and burned day and night – with such a blaze that there was no darkness in the town. Such pyres were sometimes kept burning for two or three years! Furthermore, accidental fires were extremely common, and still are in America to the extent of one hundred and fifty thousand a year. When at last this became recognized as a menace and men were engaged at high wages to fight forest fires, it only made matters worse – the number of fires increased because out-of-work men started some more in order to be paid for putting them out.

It will be seen from the above that at first the lumber trade was of less account than the actual business of clearing the ground for cropping. But the steady growth of mill-power at length made the lumber merchant very rich and powerful. When we realize that the first water-power mill in 1631 could only cut 1,000 board feet a day; that in 1767 the gang-saw cut 5,000 a day; that in 1820 the circular saw cut 40,000 a day; that in 1830 the steam saw cut 125,000 a day; and that the figure is now 1,000,000, we can appreciate what the lumber industry began to mean, and with what ruthlessness the Lumber Kings would ravage their way through the trees with an even greater recklessness than the farmers. Forests once covered six-sevenths of the State of Wisconsin with hemlock and pine, and by 1899 the lumbermen, employing 1,033 saws, were cutting 34 billion feet a year, until in 1932 there was nothing left. We may fairly call these lumber merchants tree-butchers since wood bore no more relation in their minds to the living object than slabs of meat are related to an animal in the eyes of the butcher. The trees were not trees but dollars in terms of 'timber' to be translated by the marvellous ingenuity of man into all the things, the endless things, required by civilization.

One of these things is paper. Perhaps it is when we turn from the saw mills to the paper mills of the factories in the towns that we get the clearest picture of such transformation. Let me bring the story up to date. At the time of writing this book the remarkable metamorphosis of trees into print is still in full swing and the facts concerning it are forthcoming from reliable sources of information. An average issue of the Sunday *New York Times* of ninety-two pages, plus book supplement and magazine pages, requires *one hundred acres* of forest for its production. Some American Sunday papers run to one hundred and twenty-eight pages and have a circulation of one million. This requires the pulp-wood production of *one hundred and forty* acres for each issue. Since this means the consumption, for each issue, of one thousand one hundred and twenty cords of pulp-wood, the operation demands the use of fifteen thousand, six hundred and eighty trees.

In my mind's eye, as I write these words, I can very clearly see what fourteen acres of trees mean. For it was once my job, as a woodman, to thin a wood of about that size. It seemed quite a large area to me. That a space of one hundred and forty acres should be needed for a newspaper edition, is difficult to believe; and I did not believe it until I received an authoritative and very detailed communication on the subject from the Director of the Canadian Forestry Association — and the facts are as given above. So I must accept it as a fact that every Sunday when an American family open their weekly newspaper, they are entitled to say, Here goes another fifteen thousand trees.

How can re-afforestation possibly keep pace with this?

3. The Campaign Against the Prairies

This great campaign towards the annihilation of the American forests was a very big movement involving an army of several million deforesters in one capacity or another. A movement on the grand scale must always be informed with an idea. It must have its philosophy. It must proclaim a doctrine. This was not lacking here. The philosophy of the invaders of this huge and bountiful continent was the philosophy of *inexhaustibility*. The idea was simple: there is no limit to the wealth in this country. The doctrine was pure: we are the masters and lords of this land, and may do as we please. The command was clear: pillage and pass on! There is more beyond.

This philosophy seemed particularly applicable in relation to the soil itself. Here was abundance. Here was fertility. Here in the forest clearings was loam which harboured the unpublished virtue of accumulated centuries, and had received into itself year after year the tribute of fallen leaves charged with the chemical elixirs of the air. It was natural that when they grew crops in such places, or in the ashes of forest fires, very good results were achieved. It was considered unnecessary to husband such resources. In many cases there was little thought of husbandry at all. For the first time in history fields began to be thought of as growth-factories. Step up production where such remarkable crops are possible. Repeat this again and yet again. Consider the plants as 'plants' in the industrial sense. Regard the idea of humus as humorous. Defy the laws of rotation and return. Take no notice of Nature's rule of variety, and spread monocultures over large areas. The results will be good for some time at any rate, and if eventually the plant shows diminishing returns, why then we can move on to fresh pastures, for this is the land of Space, and we are the children of Speed.

That was the general philosophy of agriculture. All the settlers cannot have subscribed to it. We may be sure that some men must have thought that they would find a quiet haven in a new world where they could live at liberty and work with

Nature, and that having built up their own little homesteads with a few chance boughs and scattered stones, they could remain in peace for evermore. There must have been some such whose voices now are silent and whose homes are buried in the dust. But that was *not* the spirit of the time, the pioneering, pushing, gold-digging spirit. There were always astute men out to do big business in a big way, and it was they who set the conditions of life in this new world. And they had their sub-philosophy of agriculture – which was dig in a spade *anywhere*. That was the idea, dig it in anywhere – regardless of whether it is on the side of a hill or at the bottom of a tree-denuded valley. And plough at speed on a vast scale wherever the blades will go.

A green girdle of protection had been thrown over the whole continent by virtue of forests and grass. The forests were cut down. Then came the plough to knife its way through the prairies, and thus tear the skin off the land in enormous patches. I do not exaggerate. The American plough has been named the Destroyer. To anyone who loves the plough, as I do who have known no greater happiness than when I handled one, the American story here is painful. It is the attitude of these lumber-jacks tearing down the trees, and of the so-called ploughmen tearing up the grass with tractors drawing batteries of ploughs, that seems to some of us in the Old World to be so abominable. The idea of feeling reverence towards the earth would have seemed to them immoral and even bordering upon the blasphemous. They were conquerors of Nature – at speed. They did not beat their swords into ploughshares; they used their plough-shares as swords with which they could rip up millions of acres into a loose soil exposed to the winds on the wide open spaces.

'Only a man harrowing clods In a slow silent walk With an old horse that stumbles and nods Half asleep as they stalk.' Their idea of a ploughman was very far away indeed from that man on the Dorset field in Thomas Hardy's poem. Their prototype was Tom Campbell, a ploughing magnate or king who, while ploughing his thousands of acres, often used five thousand gallons of petrol a day.

4. The Campaign Against the Animals

The offensive of deforestation and ploughing-up of the prairies did not exhaust their campaign against the green girdle. They employed cattle and sheep to eat it and trample it. They introduced a colossal live-stock 'industry'. If unexampled profits were to be made from the trading of wheat and timber, why not from beef and wool? To this end they must first remove all other creatures who might get in the way. They must declare war upon the empire of the animals. 'In veracious recordings,' says Mr Peattie, 'we have glimpses of deer, elk, antelope and bear, raccoon and fox, water fowl and salmon, whose profusion at the time of the white man's coming made this virgin land the richest in wild-life he had known within memory of his race. But when the white chips flew out of the first tree he assaulted, the ring of steel on living timber was the sound of doom for an immemorial order.' When the forest fires swept across thousands of acres, think what this meant in horror and suffering for countless children of the woods.

Let us pause to remember that there is nothing at all new in this extermination of wild animals by man. Europeans both today and previously have shown their prowess in this capacity. The land of Greece used to be celebrated for the abundance of its wild creatures. A recent traveller through all parts of the country is reported as stating that during all his journeys through the mountainous part of the land he saw but two pairs of partridges and one rabbit; and as he travelled past the scarred and tree-wrecked hillsides, hoping that he might at least be able to watch a flight of birds, he was disappointed even in this and was obliged to be content with one visitant only, a bird in black (gone into mourning for the rest) that croaked by the wayside, a raven. Nor have we been backward in this respect in Africa. Apart from the broad policy of killing off any animals that offer themselves as fair game, it happens that the campaign against the famous Tsetse Fly which is the scourge of man and of employed animals, has caused the authorities to order the destruction of 300,000 wild animals who are thought to be the carriers of the pest – though, as Mr Fairfield Osborn observes, if those particular animals do not serve as host, presumably other species will afford hospitality to the fly.

To take one further example, we may note how early in the day the inhabitants of Scotland set an example in this matter. I quote from Havelock Ellis, who being always at least fifty years ahead, was not discerned by the crowd behind, and indeed they hardly caught a word he said. Thus we might with advantage hear a few words now. 'Through superfluity of cleverness and wickedness – however admirable each of these qualities may be in moderation – Man has involuntarily entered into a contest with Nature, fatal alike to him and to her, yet a contest from which it is hard to draw back. Its fatality for Nature we see at every hand. As regards one small corner of the world, Dr James Ritchie, who speaks with authority, has lately drawn a terrible picture in his substantial work on *The Influence of Man on Animal Life in Scotland*. Man arrived late in Scotland – he had already touched the Neolithic stage – and he found rich fauna there on his arrival. He proceeded to destroy utterly the nobler fauna of free and beautiful creatures – many of them working for his good had he but known – and replaced it by a degraded fauna, virtually of his own creation, and yet only existing to prey upon him. He found the reindeer and the elk and the wolf and the brown bear and the lynx and the beaver and the otter and the buzzard and the bittern and the water-ousel and the golden eagle and the sea eagle and the osprey and the great auk. And he killed them all. And in their stead he placed by countless millions the rabbit and the sparrow and the earthworm and the caterpillar and the rat and the cockroach and the bug, scarcely or at all found there before he brought them, and they have flourished and preyed. For, as Ritchie has shown, Man's influence upon Nature, even when it seems but tiny and temporary, is yet in its total effect greater than imagination can grasp.'[1] And Havelock Ellis adds that in result today we see

1. Havelock Ellis, *Impressions and Comments*, 3rd series. Sorry he includes the earthworm!

man everywhere surrounded by a cloud of animal and vegetable parasitic vermin, from rodents to bacteria, multiplying at such a rate that he cannot overcome them, while they prey upon him and slay him and make him so generally poisonous that when a boatload of Europeans is landed on a remote island inhabited by simple natural men it has sometimes left death-spreading infection behind.

While it may fairly be said that we in the Old World are thorough enough in these crimes and follies, the Americans are justified in feeling that there is room for improvement in the rate of destruction. They displayed some progress in this. On reaching the great plains they found the buffaloes. Through the tall and deeply rooted grasses with their seas of flowers in spring, these animals moved in herds ten thousand strong. The trampling of their hooves could be sensed from a long way off by the trembling of the earth, while the bellowing from their throats at the mating season could sustain a comparison with thunder. Numbered by the million, there was yet pasturage for them all as they migrated through the plains and wandered by the streams and roamed over the mountains, leaving broad paths and even splendid thoroughfares in their wake. They were many; but it did not take more than a century for the whites to press them back to the wall of the Rocky Mountains and there slay them for sport, leaving their carcasses to rot on the plain after their tongues had first been cut out to provide a tasty dish for the epicures in the towns. Following the buffaloes, came the deer, the elk, the moose, and the bear to supply further targets for the rifles of the sportsmen. 'On high peaks,' says Mr Stuart Chase in a lovely image, 'were mountain goats and mountain sheep clinging with airy grace to the edges of eternity.' They also were fair game and good sport. The fun gathered speed. The sporting champions came into their own. One man claimed to have slain in a single year one hundred and thirty-nine thousand birds and animals, while another countered with the disposal of seven thousand ducks in a season, though both were considered inferior to the prowess of the Chesapeake hunters who with swivel guns killed one thousand five hundred ducks in eight hours for a New York market. Soon the sound of wild life grew faint. A silence fell upon the land. The wild-like swish of the wings of the passenger pigeons whose flight had once darkened the skies for days on end as they passed over Kentucky, was stilled, and their bodies were thrown to the pigs. The beaver built his dams no more – he had been turned into hats. The fish could no longer live in the many lakes, rivers and brooks which had been turned into sewers. The call of millions of water-fowl was not heard again when marshes were drained for unnecessary croplands. By the present century it was reported that nine species of bird and ten species of mammal were entirely extinct, while twenty-five species of bird and twenty-six species of mammal were threatened.

And the wild horses who had poured over the plains in droves? Not one roams now, even within the memory of living men. Are they then lost to us, and shall they never more be seen? Ah no, we see them still; they yet print their proud hooves on the receiving sod before our very eyes – for have they not been sung by America's own mighty Melville?

Most famous in our Western annals and Indian traditions is that of the White Steed of the Prairies; a magnificent milk-white charger, large-eyed, small-headed, bluff-chested, and with dignity of a thousand monarchs in his lofty overscorning carriage. He was the elected Xerxes of vast herds of wild horses, whose pastures in those days were only fenced by the Rocky Mountains and the Alleghanies. At their flaming head he westward trooped it like that chosen star which every evening leads on the hosts of light. The flashing cascade of his mane, the curving comet of his tail, invested him with housings more resplendent than gold or silver-beaters could have furnished him. A most imperial and archangelical apparition of that unfallen Western world, which to the eyes of the old trappers and hunters revived the glories of those primeval times when Adam walked majestic as a god, bluff-browed and fearless as this mighty steed. Whether marching amid his aides and marshals in the van of countless cohorts that endlessly streamed it over the plains, like an Ohio; or whether with his circumambient subjects browsing all around at the horizon, the White Steed gallopingly reviewed them with warm nostrils reddening through his cool milkiness; in whatever aspect he presented himself, always to the bravest Indians he was the object of trembling reverence and awe.[2]

In place of the wild life that had held the natural equilibrium, entered the employed animals, the cattle and the sheep. The grass and the bison had thrived together, but the hooves of the steers in excessive quantity punished the prairie, and the paths became gullies, while at the same time the sheep, cutting into the very heart of the grass, killed acres of it outright. Indeed, so unpopular did the sheep become in the eyes of cattle-men that the latter fought the sheep-men, call them not shepherds, until sometimes on a dark night there would be the slaughter of twelve thousand sheep. This was the true background to the romance of the cowboys of the Wild West from the Potomac to the hills of Kentucky, from Ohio to the reaches of the Mississippi, and the long rolling verdures of the prairies and the plains. The cowboys were but the tough hirelings of the Beef Barons whose business grew to such dimensions that thirty million dollars of Scotch and English capital were eventually involved in it. For we must never forget that the British policy of 'cheap food' was a main cause in the destruction of American soil.

5. Ancient versus Modern Idea of Nature

We have come a long way in this story from the conception of trees as gods. A very long way indeed. It is not a happy contrast. Could we draw back and regard these operations of our forefathers in America, with the eye, not of a human being but of an animal involved in the invasion, they would offer a fearful spectacle. Since we cannot do this we can at least try and see it as from a high place. For many centuries the land had remained under the equilibrium of the Natural Order. Suddenly it was broken in upon by a race of men from across the sea. And with what violence, with what hatred against all living things! There had never been anything like this before in the history of man

2. *Moby Dick*

and Nature. There had been many civilizations. Men had grown up with Nature in this place and that place. They had seldom been wise or good in their relations with earth. They had made many mistakes, huge blunders in tree-killing, soil-injury, and water-wastage for which they had been repaid with dust and sand. But there had never been anything like this that occurred, and occurs, in modern days under the sign of mechanism. When we think of the ripping up of the grass in every direction, of the crashing down of huge trees under the axe and the deliberately lit forest fires rushing forward at the rate of an express train while thousands of animals shriekingly fled in terror from the crackling flames till exhausted they were burnt alive, our minds and our hearts turn back to the primal days of religion and reverence. For what had those beauty-blind mechanical destroyers, those reality-scared lumberjacks, to do with any sort of religion? Do they not seem only as large insects, or as a plague of locusts eating everything before them without the excuse of being driven blindfold in the coils of necessity? We think back, I say, we turn our hearts again to the days when man was unclouded by his comforts and uncorrupted by his engines. We think of the origin of ploughing and the first mystery of milk, of the meaning of the cow and the mourning at harvest.

Consider the ancient fields. Men hardly dared touch them. The first plough, made of gold and drawn by sacred animals, was used only as a ritual in the temple gardens to do homage to the god of fertility. Even so, that scratching of the soil was regarded as a kind of *wounding*, a sacrifice of earth for the good of her children. The plough was considered so sacred that the fugitive who reached it found protection as certainly as the thief who stole it suffered retribution. The idea of being torn – or ploughed – was linked with the mystery of generation, and the Indian earth-mother Prithivi, having first surrendered herself in the form of a cow to Prithu, could only become fruitful when her lap was torn as with a plough. This symbolism was also recognized by the Athenians who legally sealed Wedlock upon the implement; and the ancient Irish, embracing in one mighty image the starry heavens and the fruitful earth, first spoke of the constellation of 'The Plough'.

Thus also with oxen. The origin of oxen as partners with man was equally rooted in ritual. The sacred bulls that first drew the holy chariot in the procession of the temples were so sacrosanct that the killing of an ox was an act of murder punishable with death; and even after the annual ox-sacrifice at the Athenian rite of Bouphonia, the presiding priest was compelled to fly from the country and to throw his bloody axe into the sea.

Thus also with milk. It was unthinkable as nourishment and conceived only in the light of a ceremonial drink, and so treated by the Orphic and Dionysian priests that it induced a kind of spiritual intoxication, its deathless properties being symbolized by the Egyptians in many a mysterious rite. Thus we see representations of the goddess Nuit in the form of a cow suckling the King to bestow upon him the gift of eternal youth; and just as in Egyptian wall-painting the plough is the tool in the hand of Osiris, so we can read on their tomb-inscriptions 'Isis giveth thee milk', or 'The Cow Hesah giveth thee milk'. In what lay its virtue? Do we know and can we tell that mystery? Or are we far too much enfolded within layer on layer of crusted knowledge that divides us from the pristine impact on an unscored mind? Perhaps it had something to do with its *whiteness* – the colour that in all ages has served as the symbol of divine spotlessness and power. Certainly we may say that if the ancient nobility of Scotland were washed with milk, and the Picts of Ireland used it as a salve against poisoned arrows; if Romulus poured out a libation of milk at the founding of Rome, and the Todas of India laid their dead on a bier in the milk-chamber; if the milk of human kindness was numbered by the early Christians as the greatest of all virtues, and a never-empty pitcher of milk was lifted to the lips of the martyrs; if to the still sensitive souls of men there is no loftier sign that the Milky Way, we must conclude that here we have a real feeling for the profundities of existence not unmixed with dread at offending the demands of deity.

This approach to the Creation did not last. The day came when the ceremonies fell into disrepute and a 'practical' application was made to the ritualistic object. Then the plough could become a tool. The ox could become a carrier of burdens. Milk could become nourishment. A rent was torn in the mind of man and we entered the Era of Economics. In order that milk might become food the symbolism had to be forgotten. 'In effect the spiritual world had to be denied,' said Dr Pfeiffer. 'What a mighty fall of humanity took place when the holy emblem of deathlessness, of the passing of the threshold at birth and death, became mere nutriment.'[3] Is this the real Fall of Man, or at least for us today, the Fall that means most? Is it this that creates the unease which shadows us always as we walk in the harsh light of modernity?

When we remember these early conceptions regarding the earth, and the wounding of the earth, and the sacred oxen and the ritualistic milk, and then turn to the Beef Barons and the Ploughmen Kings and the Timber Magnates of America, it is hard to equate them with progress. Wounding the earth! – with what axe could one open the skull of a modern lumberjack on the spot or lumber merchant off the spot, and plant that idea therein? It is totally beyond the conception of the big tough guy with the tiny mind. Think of the modern harvest with its hundredweight sacks fetching so many dollars or shillings, as against the gathering and the garnering in the ancient days amidst weeping and voices raised in mourning songs! They felt guilty for what they had done. They had robbed the earth, their mother. They had taken without giving. They must show contrition and entreat forgiveness. Amidst the lamentations of the populace an image of the goddess, Demeter, in the form of a mighty grape, must be carried in triumph and then dismembered as a symbol of the bleeding vine; and just as the primitive African apologized to a tree he had cut down or injured, so must the disciple of Manes at the harvesting of grain entreat the presiding deity: 'Not I have harvested thee, nor have I finely ground thee, nor do I place thee in the oven. Another accomplished this deed. Guiltless I ate it.'

3. See Ehrenfried Pfeiffer, *The Earth's Face and Human Destiny*, chap. 9, to which the present writer is deeply indebted.

Such feelings were widely experienced. From the Sagas of the north to the Mysteries of the south, from the ballads that celebrated the torment of John Barleycorn to the mourning maidens of Islam who at the baking of bread bewailed the passion of the god, Tamuz, we see in many forms and under many signs the expression of guilt, the lamentation for death, and the belief in resurrection. It was not until these ideas faded that the harvest could become an occasion for gaiety, festival, and riotous rejoicings. Such scenes would have appeared as shocking to the primitive ploughman as our gross commercialism of Christmas would seem to the early Christian. Today there is neither weeping nor rejoicing. We are not sorry, and we are not glad. We have no time, for we must keep pace with the machines. Even the gracious ceremony of grace before meals has been abandoned; for that would seem too much like thanking Tom Campbell for ploughing those acres for which he has already received his pieces of silver. We owe debts to no one. We have nothing to fear.

6. Again the Trees Reply

Or have we something to fear? Until recently we did not think so, and the Americans certainly did not think so. But now Nature has made her reply and shown her hand. Let us watch what happened in the North American continent – or, rather, some of what happened in some places, for as I have said before, I am concerned in this book to make a statement, not to give an exhaustive survey.[4]

Let us proceed with caution. I am anxious to avoid exaggeration in this matter. There is no need for rhetorical tropes. Every American was not then and is not now a villainous despoiler of Nature. Furthermore, if we permit our minds to move freely in contemplation of earthly phenomena, human and otherwise, we cannot fail to recognize that the planting in a few

4. For the alarming story of South America, see Vogt's *The Road to Survival*.

years of a civilization on that mighty virgin land, is one of the marvels of history. The sowers sowed the seeds of cities, roads, railways, machines, factories, men (above 150 million). It may be that they will not last longer than flowers, and will sink back into the spent and unconsidered earth. But how prodigious the effort, and how magnificent the show. When we approach Manhattan from the ocean and see those towers rising from the water to assault the clouds; and when later we walk through the canyons of New York confusing top-storey lights with stars, we cannot but acclaim the lyric splendour of that stage. It may not be permanent, and the scene-shifter behind may be compelled to remove it or its inhabitants; but that will not lessen the brief glory of the play.

At the moment that civilization is still holding up large portions of Europe as well as itself. Let us see just what it has to stand upon now. In 1630 the land offered 820 million acres of forest, 600 million acres of grassland, and 430 million acres of open woodland. Today it is calculated that not more than one-tenth of the forest remains and that the annual loss exceeds the annual growth by over fifty per cent; and it is calculated with regard to the soil that one-half of the fertility of the continent has been dissipated. That is to say that though there has been great loss of soil there is still a great acreage remaining. For we must recognize that a continent of that size whose soil has been built up through millennia cannot be utterly destroyed for some time, however enlightened and progressively mechanical the attackers may be. Nevertheless what has happened is sufficiently impressive.

The first thing, as we have seen, was the attack on the trees. They came down. They were very valuable when alive and standing, for again as I have shown in proper place, they were grips, they were stakes, they were sponges – the twigs, leaves, rotting logs, pebbles, and stones at their feet serving as a filter and retarder of floods. They were mowed down and the ground was ploughed – that is loosened – where they had stood. That is the first thing. Then the tiny trees came down. I mean the grasses, for in relation to the soil they might be called little trees, as they also keep it in place by their root-grips and wind-cushioning stems. They also were attacked, mown down and ripped up by speedy mechanical means, and prairie fires were lit that rivalled the conflagrations of the woods. Those were the two main movements in the subduing of the wilderness.

And they were the two main causes in the erosions that followed. At the very moment when the pegs were taken up the carpet was cut into pieces. It was exposed to wind which first having dried it could then blow it away. At other places, exposed to unchecked torrents of rain, it could be washed away. The details concerning this have been rehearsed over and over again by the experts on the subject, and therefore brevity here on my part is surely desirable. We need but to remember a few salient facts. Nitrogen, phosphorus, potassium, and sulphur, are four of the most important ingredients in the soil. In a virgin land they have been built into it through centuries of undisturbed give and take in the diurnal round. They are so light and insubstantial as to be defenceless if exposed to wind or flood. Thus bare ground turned by the plough can lose as much fertility in ten years as unbroken prairie in four thousand years (which also forms even as it goes). Consider the irreplaceable network of filter and sieve which we call roots. Remove them and the result is that water simply messes up the clay and the sun bakes it into chunks which are shifted into the valley by the next shower if the land is on a slope – while the unpercolating water rushes as from a roof, carving a gully in its passage. On a steep field a pouring thunderstorm will let loose up to four hundred tons of water per acre per hour – all of which could be successfully channelled if cushioned by trees. Without resistance four hundred tons of water can make itself felt and leave its mark. A piece of land in Georgia which was forested and calculated as having lasted for 35 thousand years, was uncovered and showed signs of disappearing in 25 years.

Consider also the effect of vast monocultures over wide expanses of cropland – of course those minerals were used up whether they were blown away and washed away or not. They went out over the railways into the cities and across the seas. It is so easy to forget that we eat the soil. We in Europe have eaten a great deal of American soil. They have recklessly shipped it across to us. It has been sent in the form of wheat or beef – the latter having first over-grazed and hoofed the soil it fed on.

Once more, think of the inheritance of the new people in the New World. They came to two billion acres, half in forest; forty per cent in strong grasses; only two per cent in desert. From the Atlantic to well beyond the Mississippi stretched unbroken primeval forest. That was the wealth they took over, the deposit they found in the earthly bank. They broke into the chest and rifled its contents, calling their action 'sturdy individualism' or 'ameliorative improvements', or simply 'the enterprise of capitalism'. What is the opposite to that kind of capitalism. It is not socialism. It is not communism. It is conservatism. The opposite to capitalism is conservatism. But up till this century the idea of conserving anything never entered the American mind. The fantastic towers of the speediest growth in history rested upon the swift plundering of Nature's hoarded wealth. That was the foundation upon which they built their house. Was it built on sand? And is it writ in water?

At first the reckoning was not easy to discern. Good yields were naturally expected and achieved at the beginning – it was a long time before a depreciation of *fifty per cent* was discovered! It is not easy for farmers to discern the least spectacular but most deadly form of soil wastage, namely sheet-erosion – the invisible washing away of the surface good stuff in high places into the valleys and the rivers. The farmers above who began to lose in their yields did not know what was happening any more than the farmers below who became richer by virtue of the unexpected deposit. The latter began to wake up and to become less pleased when later on the offering they received from above was in the form of useless under-soil and then of pebbles and stones.

That was sheet-erosion, to be found wherever the land sloped – and eighty per cent of America is sloping. But manufactured erosion is dynamic and cumulative. It grows by what it feeds on. In the civilization of speed it was itself speed. At last the

Americans began to look round with alarm. They had taken over the soil. But it would not stay still, it was not a stable thing. They had cut down the trees. They could not cut down the wind, they could not control the rain. The accumulation of twenty thousand years of soil-building was failing before their eyes – in a single century thousands of years of fertility was being thrown away. There is a belief that we live by the soil. There is a general idea that we live on mother earth, and since there is plenty of earth, all is well. The trouble is we only live on the *top* of the earth, we have only the *top* of the soil to play with – a spade's depth. Thus are we perched precariously. If that surface goes we go with it. That surface began to go rapidly in America, as elsewhere, and men to go with it.

After the sheet, followed the gully erosion. Small gullies carved by torrents unarrested by any obstacle were so deepened and widened that in some places spectacles of astonishing bleakness and devastation were created. Canyons had literally been carved in less than a century over hundreds of acres which previously had shown neither gully, nor ravine, nor canyon. That was bad enough, but the increasingly uncontrolled nature of the rivers was more immediately catastrophic. Living by the banks of a great river began to become more dangerous than living under the shadow of a volcano. Some of the towns that had been built upon the banks of the Mississippi were no longer safe, and at a moment's notice could be and often were submerged in flood or totally swept away. Yet this loss to human life and property was nothing compared with the continual day by day loss of soil carried to the sea. The Mississippi alone carried – and still carries – 400 million tons of solid earth to the Gulf of Mexico, taking with it the microscopic organisms that make humus what it is – the minerals amounting to 40 million tons of phosphorus, potassium, nitrogen, and sulphur. Already by 1936 it was calculated that 100 million acres of formerly cultivated land had been essentially ruined by water erosion, which is an area equal to Illinois and Ohio, North Carolina and Maryland combined; it was further calculated that the washing of sloping fields elsewhere accounted for the stripping of the greater part of the top-soil over another 125 million acres; and that further, calculating from approximately ten years ago, another 100 million are on the way out – which gives a total of over 300 million acres likely to be confined to the dust-heap.

It will not amuse any American to speak figuratively of dust-heaps. For that brings us to the worst form of erosion. The carriage of the soil to the sea by water is not greater than carriage by air. The Great Plains were opened to the winds, and the wind has been carrying them away ever since. This form of erosion, taken with the action of water as above, brings the figure up to three billion tons of top-soil lost every year. A storm of dust on 11 May 1934 meant that 300 million tons of fertile soil was swept off the great wheat plains and carried to places where it was utterly useless. It meant the laying down of sand dunes and the creation of landscapes that look (in photographs) as if an enormous snowstorm had covered up all but the tops of houses, trees, and ravines. Everyone knows that the now famous Dust Bowl which takes in portions of Oklahoma and Kansas, Colorado and Texas and Wyoming, blows its farms two thousand miles away into the Atlantic Ocean.

These storms do not dispose of more soil than is accomplished through the agency of water, but by all accounts they are unspeakably vile. To die of 'dust pneumonia' in that once clear-lit land upon which nature smiled with bountiful promise, must be a terribly unpoetical end. To see your farm submerged in dust, as has happened, must promote feelings far more drab than those of the Etna peasant whose farm may disappear, not owing to man's folly, but to the fiery nature of the earthly depths. In the year 1935, in a derelict and wasted farm between Wyoming and Texas, the corpse of a man was discovered in the sand. What had happened to him we wonder? Had he been ill and unable to escape? What secret lay buried in his tomb of dust? And what did the dying eyes behold? A height had loomed up behind him. A mountain? Ah no, it was not a solid: no crags were there, nor crooked cliffs upon which the eye could rest serenely. It was not a liquid: it had no vapour in it that might let fall rain to drench his parched and wasting land.

It was the soil itself that had got up there. It was America blowing away. 'Behold my country!' he may have cried. 'Behold our dreams, our hopes, our plans and works, our triumphs and our spoils – now turned to dust!' And perhaps in those last lonely hours he may have fully felt the awful folly and the vast rebuke.

7. The Refugees from the Wrath that Came

The scenes of dereliction were not confined to abandoned fields and prairies turned to dust. The ravages of the revenge were not marked only by hideous gaps and gullied scars on land where once the trees had stood. We see another scene as well – the spectacle of deserted houses, deteriorating farms, and a horde of hungry, wasted, angry men.

Remember the philosophy of the men of speed. The land is inexhaustible: get what you can here, there is more beyond. It was a sort of mining above ground. They would rush at a place, cut down all the trees, set up a great lumber industry, build a town, and promote a thriving agriculture in the clear spaces. Then one day they would begin to find that there were no more trees left in that area, and that the soil was not suitable for concentrated agriculture after all! The mills would fall silent – the mills that had ground so exceeding fast. This led in many cases to the abandonment of whole towns and even of whole counties.

Thus they arrived in the Jackson County of the Lake Superior region. The lumber men assailed the white pine. The trees fell down. Mills, railroads, stores, houses took their place, and the region boomed. Then the farmers moved in to share the prosperity. The pine having been annihilated and exchanged for the pleasing spectacle of stump, scrub, and brush, the ex-timber workers cleared more land for farms, and put into operation big drainage schemes. It turned out to be too sandy. There was complete failure for farmers on the spot. But in the cities behind, it all provided a speculator's holiday. They began to promise fortunes out there to unsuspecting townsmen. 'Come to Jackson County and make your Fortune in its Rich Irrigated Loams.' The suckers sucked. They came, they saw, they failed, they starved. The whole county went into bankruptcy, and streams of *refugees* trailed across the land at the mercy of the successful, strong settlers who had gone ahead and captured further land. These were the grapes of wrath; these were the men fleeing from the wrath that had come in the wake of folly and greed. And there were many of them, multitudes three thousand strong swarming on the highways. They had come out to exploit the land and had been exploited by the first exploiters. They would gladly have defaced the earth for cash, but others had got in before them, and nothing now could be reaped as nothing sown save a crop of infamy. Hating and hated, bitter and violent, these the foolish ones turned their beauty-blind eyes with devouring greediness upon the possessions of the wise, and as they flowed westward, wave upon wave of them hungry, and homeless, through Kansas and Arkansas, Oklahoma and Texas, New Mexico and Nevada, they presented a spectacle of defeat and shame, of misery and degradation such as we associate with the refugees displaced by the wars of man with man. In this case the spectacle was one episode in man's conquest of Nature – when the conquered were the conquerors.

It may be urged, of course, that we must not take the plight of these people too seriously. They represent a mere three hundred thousand out of one hundred and fifty million. A sad chapter, but not important. It is over now. It cannot happen again since all the land is plotted out. Yet is it over? Is the hour of famine, is the sight of refugees fleeing from the wrath of outraged Nature, a thing of the past? Voices are still raised in warning that deserts in America, *equal to the size of the Sahara,* must follow the depletion of soil which Nature can only replace at the rate of an inch every five hundred to a thousand years. It is said that in forty years there will be no reserves of timber; and there are not lacking authorities, authoritative or otherwise, to declare that in a hundred years there will be no food resources left. For myself, I have too much faith in man's respect for his own skin and in the present conservation schemes, to think that it will come to this. But how easy to fail in vigilance! How tempting to throw aside caution in favour of short policy and quick business. The measures exercised by the farmers who saw the disaster in the Thirties have not yet become the acknowledged practice, while the Second World War, by offering high prices for crops and meat, caused the very same neglect of conservational principles which before had brought about the Dust Bowl. Recently, in Texas alone, according to Mr Louis P. Merril, it is estimated that more than one million five hundred thousand acres of land unsuitable for cultivation have been ploughed. The result of this sort of thing is that in the drought of 1953 approximately five million acres of land suffered moderate to severe wind erosion in the southern Great Plains.[5] This looks like 1935 again – another Dust Bowl? This has been greeted with justified alarm – and by some most unjustified surprise. When there is less surprise there will be less cause for alarm; but it is clear that before America comes to terms with Nature – if she ever does – we will see a great deal more punishment exacted in unyielding soil and rotting men.

When I think of these things, my mind goes back to certain scenes in an Englishman's book. The author was a man of transcendent genius, whose ability to present atmosphere, amongst other things, was so great that his works are read in every portion of the world where books are known. Near the end of the nineteenth century he paid a visit to America, and when he came back he wrote *Martin Chuzzlewit* – in which he sent his hero to that country. When I read the book for the first time I was puzzled by the account of Martin's dinner with Mr Jefferson Brick and others. A bell rang. Instantly the gentlemen who were walking with him dashed up some steps into a street door like lunatics. Evidently it was an alarm bell and the prem-

5. 'This is comparable to the amount of land damaged in 1935, and about twice as much as was damaged in any one year since 1938. Some sections are in worse condition than at any time during the Thirties' (William M. Blair, *New York Times*, 4 August 1953).

ises were on fire. While Martin was looking for the smoke, more of the company, with horror and agitation depicted on their faces, came plunging wildly round the street corner, and all together threw themselves in at the same door. Entering to see what was up, Martin 'was thrust aside, and passed, by two more gentlemen, stark mad, as it appeared, with fierce excitement.' When Martin was in the room he could see no fire, but only a dining-room in which a meal had just been served. 'It was a numerous company,' says Dickens, 'eighteen or twenty perhaps. Of these some five or six were ladies, who sat wedged together in a little phalanx by themselves. All the knives and forks were working away at a rate that was quite alarming; very few words were spoken; and everybody seemed to eat his utmost in self-defence, as if a famine were expected to set in before breakfast time tomorrow morning, and it had become high time to assert the first law of Nature. The poultry, which may perhaps be considered to have formed the staple of entertainment – for there was a turkey at the top, a pair of ducks at the bottom, and two fowls in the middle – disappeared as if every bird had the use of its wings, and had flown in desperation down a human throat. The oysters, stewed and pickled, leaped from their capacious reservoirs, and slid by scores into the mouths of the assembly. The sharpest pickles vanished, whole cucumbers at once, like sugar-plums, and no man winked his eye. Great heaps of indigestible matter melted away as ice before the sun. It was a solemn and an awful thing to see. Dyspeptic individuals bolted their food in wedges; feeding not themselves but broods of nightmares, who were continually standing at livery within them. Spare men, with lank and rigid cheeks, came out unsatisfied from the destruction of heavy ditches, and gazed with watchful eyes upon the pastry. What Mrs Pawkins felt each day at dinner time is hidden from all human knowledge. But she had one comfort. It was very soon over.'

Why were they so hungry? I wondered when I first read that description. Why were they so continuously hungry, according to Mrs Pawkins, 'as if a famine were expected to set in before breakfast.' The passage has the force of unconscious symbolism or prophecy. Reading it today it seems like a picture in miniature, either of the manner in which Americans were busy devouring their continent, or of men frantically feeding before the hour, close at hand, when there would be nothing left to eat. It is strange that Dickens, a nineteenth-century Londoner, who knew less about forestry and soil erosion and cared no more about them than Bill Sikes himself, should have made a fable here. But then Dickens was Dickens, who had that enormous capacity for sensing the essential in an atmosphere. When I took up *Martin Chuzzlewit* for the first time, I remembered looking forward to reading his famous attack on America – scenes, I supposed, of Dickensian humour playing upon American foibles in the land of boundless promise and prosperity, as it then seemed. But no such thing. That was not the impression – or not chiefly. The main impression left in the reader's mind is the ironically named Eden where Martin and Mark Tapley go to seek their fortune – a ghastly place in the wilderness, a petrified forest, a swampy sink of fever. The easy victim of some speculators, Martin went to just such a place as thousands of other optimistic land-greedy men arrived at in those days, full of hope which quickly changed to despair.

Those two main impressions made in the mind of a reader of the American part of *Martin Chuzzlewit* are the very last things which one would expect Dickens to hold up at the end of the nineteenth century in the land of apparently boundless activity, opportunity, and prosperity. Yet such are the pictures that the great man did hold up – famished diners, and 'cities' of dereliction and dismay.

The Americans had declared war on Nature, and had come away with many spoils and triumphs. Then Nature declared war on them; already by 1936 when *red snow* fell in New England, it could be calculated that the dust offensive and the water offensive had taken away one-half of the original fertility of the continent; and we are now assured by the Soil Association that its annual loss of productive soil by erosion is 3,000,000,000,000 tons – enough to fill a train of freight cars girdling the earth eighteen times, and the equivalent of 73,000 forty-acre farms washed or blown away.

The trees had fallen. Nine-tenths of the trees had fallen from all that mighty host. They were dead, and since their carcasses are much more useful than dead men, they were marvellously transformed into a thousand implements. But suddenly it was found that they were even more useful when alive. They were the only police force that could protect man himself from the ravages of tempest and of flood. God was, after all, on the side of the green battalions, and man must retreat before the mockery of this Moscow.

PART FOUR
CONCLUSION

Man is a dramatic animal. He leans towards tragedy. He courts disaster. Born finally to obey or perish, he must first defy the laws of earth and usurp the throne of God.

We have followed something of this drama. We cannot pursue it to the end on paper, for the last act has not yet been played. The climax may have been reached but not the dénouement. There is another act to go. Since we are the players, what happens depends on us.

When trees were regarded as gods they were not cut down. Hence the mountains also stayed up, and the soil remained steady, and the waters true. When trees came to be regarded simply as 'timber' they were ruthlessly slain. We have rehearsed some of Nature's replies with the swollen river and the plague of rodents, with the day of drought and the bowl of dust.

Yet there is this to be said about man. He learns from history. History is experience. When we put our hands in the fire, we learn never to do that again. If the experience is painful we learn and we act. Already the Americans are learning. They are taking steps, if still haltingly. Having nearly destroyed themselves by their exertions they may save Europe by their present example. They are soil-erosion conscious as no nation has ever been before. They know what has happened. They know what they ought to do. They have experienced a sufficient amount of calamity to make them act. Man must have calamity, he must have disaster, before he can save himself. The Americans saw the floods rise, and the dust blow, and the earth melt from beneath their feet. When such things happen men become sincere. For today, as at all times, the human animal, nearly shipwrecked, will turn towards some means to save itself. The ideas of the shipwrecked are the only genuine ideas, said Ortega y Gasset: all the rest is rhetoric, posturing, farce. The shipwrecked man 'will look round for something to which to cling, and that tragic, ruthless glance, absolutely sincere, because it is a question of his salvation, will cause him to bring order into the chaos of his life'.[1] The Americans heard the cries of the shipwrecked – and they have acted. The whole world has heard of the Tennessee Valley Authority. What they did there, what they are doing elsewhere in the replanting of their forests and the conservation of their soil, may save them from total disaster and eclipse.

The story of man and trees in Western Europe has not been so calamitous in result as the American. Nevertheless nearly a thousand years' war was declared against trees in Western Europe. It is not the fault of the deforesters that the land is not now in worse shape than it is – the fact is of course that the invaders of the American continent had the conditions for a real rape of the earth, not hitherto available. The European situation with regard to deforestation is too complicated to be subject to a general statement. But it is only too obvious that today, the Scandinavians, who have largely kept their trees, are in a very much better position with regard to fertility than Southern Europe, especially Spain, whose once tree-covered country is now a scene of almost Eastern poverty. Again, the Germans are world famous for their forests and their foresters. Today the trees are falling fast under the axe of the occupying powers. But let us remember that it was the Nazis who first set about exploiting German and Austrian forests without any regenerative policy whatever. They wished to use their trees as weapons of war, and did so with such thoroughness that by 1942 it could be written:

Clad in fabrics produced from wood, living on wood sugar, wood proteins, and meat and cheese from wood-fed cattle, with a shnapps ration made from 'grain' alcohol obtained from sawdust, German soldiers move to the Russian battle lines in wood-gas-driven trucks, which are greased with tree-stump lubricants and run on Buna tyres made from wood alcohol. Spreading misery and destruction with explosives manufactured from the waste liquors of woodpulp mills,

1. *The Revolt of the Masses.*

they are assisted in their nefarious work by squadrons of plywood planes, while the German propaganda division takes a motion-picture record of selected items of the action on a film made of wood cellulose acetate.[2]

The trees have had their revenge all right. And, one way or another, it will return on us, we may be sure, if we further deface the German forests.

In England forests once covered nearly the whole land. Envoys returned to Caesar saying that they could not penetrate to the end of them. In due course they also were cut down. There has been much re-growth since, and reckless cutting of the re-growths, but still many woods remain to the tree-loving British, while of course a supply of rainfall has never been a problem for the islanders.

For my part, I think that the danger to England caused by the primary destruction of her forests, goes very deep. The nemesis is very real and very terrible. It goes underground. Already by the fifteenth century so many forests had been cut down that wood as fuel was beginning to become scarce. When Aeneas Sylvius, later Pope Pius II, paid a visit to England in 1458 he noted in his diary how pleased the poor people were when they were given stones for alms. 'Nam paupers penenudos ad templa mendicantes,' he wrote, 'acceptis lapidibus eleemosynae gratia datis, laetos abisse conspeximus; id genus lapidis, sive sulphurea, sive alia pingui materia praeditum, pro ligno quo regio nuda est, comburitur' – the sense of which would read: 'Now we have seen begging at the temples, poor people almost naked: who, when they had been given stones for alms, went away happy. That kind of rock, which may contain sulphur or some rich material, is burned instead of firewood when the district is bare.' That is to say already coal had been discovered, and that branch of forestry which we call coal mining, had begun. First we cut down the forests standing above ground. When they were exhausted there remained the woods underground – the carboniferous forests. Coal mining is a branch of forestry and

2. 'The Rediscovery of Wood,' *American Forests*, September 1942.

agriculture; but we dig deeper, we cut without planting, we reap where we have not sown.

Thus that great day came when the carboniferous forests were located and the properties of coal were realized. Perhaps this was the most exciting discovery of all. We are weary of such things now. Our hearts are cold and cowed. But we shall be lacking in imagination if we cannot realize what it must have seemed like in those days, the excitement which the words of George Stephenson must have held for all who heard them: 'We are living in an age when the pent-up rays of that sun which shone upon the great Carboniferous Forests of past ages, are being liberated to set in motion our mills and factories, to carry us with great rapidity over the earth's surface, and to propel our fleets, regardless of wind and tide with unerring regularity over the ocean.'

We must allow a certain epic grandeur in their theme. The power was divined. The wealth was realized. The possibilities seemed boundless. Naturally there was a coal rush. Claims were staked out by the enterprising and adventurous, and messengers were sent down into the primitive forests. A strange journey indeed! Strange wanderings in those sunken lands! Pioneering down into the darkness, the travellers explored that green old world of long ago. They made perpendicular roads and descended as far as three miles into the buried woods. They carved out galleries within them. They ran trucks through tunnels chiselled from the petrified leavings of the rotten reeds. And as they passed along those corridors encased by the corrupted ferns, and penetrated even further into the lost regions of the sunlit lands, the danger from gases obliged them to go in darkness with nothing to lighten their way save the phosphorescent gleam from dried fish.

Eventually the Davy Lamp was invented; but until then they had to rely upon phosphorescent fish. A haddock, for instance, can be so luminous that it could be photographed by its own light. 'Burning without fire these lamps will not call forth force' I wrote in *Paths of Light* under the head of Phosphorescence.

CONCLUSION

It is a strange thought. It conjures up mystery upon mystery. In a coal mine are crouching forces that would leap forward in answer to a flame. Silence; stillness; the blackness of dawnless night; nothing living, nothing moving – such is a corridor in a mine. No life is there – but force is sleeping there: animation is suspended in those inky halls. The entrapped sunshine is there – imprisoned for three hundred million years. A little flame will be enough to voice its presence. Strike a match – and suddenly that crunched and crouching power will leap out like a wild beast and rend the cage. So the early visitors to these lost forests of yesteryear were careful not to come with fire that would call forth the spirit of the ancient sun. The ineffectual flame they brought was cold and tame – a fireless torch upon a lifeless fish.

They encountered more perils than explosive gases. In making their way through the subterranean forests they sometimes came upon tree-trunks standing erect, the interior being sandstone and the bark converted into coal, so that as soon as the stance of such trees was weakened they often suddenly fell, killing the men below. 'It is strange to reflect,' says Sir Charles Lyell, 'how many thousands of these trees fell originally in their native forests, in obedience to the law of gravity, and how the few which continued to stand erect, obeying, after myriads of ages, the same force, are cast down to immolate their human victims.' But nothing daunts the spirit of man. In heaving out these precious rocks, this bottled energy, for expansion in the upper world, no effort was too much, no sacrifice in flesh and tears too great, and hundreds, even thousands of these visitors to the ancient woods gave up their lives and lay down eternally entombed amidst the sepulchres of the trees.

This enterprise was pursued with such zeal and concentrated industry that during the nineteenth century England cut out more of these forests, this coal, than was cut out elsewhere over the *whole world*. This changed England utterly. Her history was altered. She was forced to enter on a road all unforeseen. It caused the colossal industrialism of the country. That is by far the most important effect which trees have had upon England. They had been sleeping below. They were disturbed. When they were carried to the surface they were in the form of great potential activity. Once they had got to work they changed everything, including characters and faces. As for their effect upon population, a twenty million increase is an underestimation. Thus England became one of the most powerful countries in the world. Then the most vulnerable. And now?

Everyone knows her dilemma now. No country in the world, or in history, has ever been less ecologically sound than the England of today with its population of fifty million, ninety per cent of whom work at non-agricultural activities caused by the carboniferous forests. That is the fact. When food fails presently to come in from other countries how will the fifty millions get on? The question is enough to make a tree laugh. This ever hanging threat is the cause of the gloom which has fallen upon the English of late – they feel they have no *footing* on earth.

The realization that things had taken a dangerous course was expressed in the phrase 'back to the land'. But it was soon laughed out of court as pleasing romanticism. The British decided to ignore this gigantic threat. The clear-eyed prophecies of woe carried through from Cobbett to Carlyle and from Ruskin to Morris were stifled by the Fabians who, suddenly and calmly *accepted industrialism*. Their influence in this was enormous. From then onwards the political sociologist could speak solely in terms of the *rights* of man instead of his *needs*, and of his welfare instead of his responsibility to earth if he is to exist at all. A whole generation came under that influence. The only thing necessary, it was said, is good distribution: distribute as many things as possible to as many people as possible and all will be well – that was the concern of economics. And even when at last Laments for Economics were raised, and the Death of Economic Man announced, there was not the faintest attempt from that quarter to make way for the arrival of Ecological Man. It was a world of plenty, we were told, and the crime was poverty in the midst of plenty – and soon, so very soon, we were to find that the opposite was the truth, and that all we can hope to see, as demonstrated by black markets, is plenty in the midst of poverty.

Under the same influence the educationalists went on and on and on declaring that what was necessary was more schools and better buildings and more teachers, while they continued to allow generation after generation of school children to grow up with the idea that manual work is degrading, that muck-heaps are dirty, that harvesting is something to do 'on your holidays', and that it is more dignified to be a nonentity in a town than a solid workman in the fields upon whom the whole world depends – the contemptible fatuity of which ideas can only be really felt by someone who knows both worlds. I will add this only here – that those ideas are not only contemptible, they strike against the psychological and physiological needs of many millions of children later doomed to a life of nonsense which they are forced to accept. And on the intellectual side, is it worth pausing to inquire before closing this paragraph, to what extent Ecology is taught?

The more influential of the *literati* took the same industrial line. The man to be respected was the workman in the towns. The middle-class intelligentsia suddenly found enormous merit in being an industrial worker. As for love of Nature, or appreciation of her laws, that was regarded as outmoded and sentimental. Instead of examining the ground upon which they stood, they gazed steadily at the Horizon. Instead of seeing the mud in the Yellow River, they dreamed of the Moon in the Yellow River. It was a conspicuously urban poet residing in London who led a generation to turn their attention from the waste land of the soil to the private Waste Land of the soul. It was he who declared that 'April is the cruellest month'. It was all very well for one great man in one bad mood to say that once. But it was not a good thing when a whole generation took this as the proper response to Nature. And yet, industrialism had reached such a pitch that perhaps it was inevitable, and there was at least sincerity in that response. It is said that in France and in Central Europe during the last two centuries, fifty per cent of the philosophers and poets stemmed from the mountainous regions. Perhaps it was natural that at a time when England had reached a culmination of industrialism and was busy ruining her farmers, April should have appeared the cruellest month to a generation of metropolitan poets whose sole connection with agriculture was confined to the sowing of a little wild oats, and who had never climbed a mountain higher than Parliament Hill.

It seems to me that the time has come for Advanced Guards – philosophic, educational, poetic, scientific – to cohere for once and make their countrymen conscious of the ecological situation. It is a comprehensive theme. 'We have learned to see in mythology,' says Dr Pfeiffer, 'a good deal of physiology and natural scientific wisdom.' In this book I have tried to bring together the intuitions of the past with the factual knowledge of the present. We have reached a time when we can get our bearings. We can discard superstition without replacing it with irreverence. We can sense the invisibilities on a higher plane of apprehension. Edward Carpenter said that he once managed to glimpse at any rate a partial vision of a tree.

It was a beech, standing somewhat isolated, and still leafless in quite early spring. Suddenly I was aware of its skyward-reaching arms and upturned fingertips, as if some vivid life (or electricity) was streaming through them into the spaces of Heaven, and of its roots plunged in the earth and drawing the same energies from below. The day was quite still and there was no movement in the branches, but in that moment the tree was no longer a separate or separable organism, but a vast being ramifying far into space, sharing and uniting the life of earth and sky, and full of a most amazing activity.[3]

We cannot all reach these visionary heights, nor can any man remain there. But we can all be ecologists. There are in America today agriculturists with astonishing practical genius combined with comprehensive ecological insight. They will pull America through *if* given the chance. And if in England similar leaders of the field are supported and allowed to lead the way and show the means towards the greatest compromise England has ever been called upon to make, the compromise between industry and agriculture, then England could regain her balance. But she

3. *Pagan and Christian Creeds.*

must make up her mind about it. The English can do anything if they make up their minds upon a course of action – but they do not like doing so, they would rather drift. Can we afford to drift any longer, in any country? Certainly if the present un-ecological life in England is continued and other countries are relied upon to support her – why, then that cutting down of the forests which led to the cutting out of the squashed and hoarded wealth of wood below, will have meant disaster. For trees always have the last word.

Rather we should say, more broadly – Nature always has the last word. And having said that, should we not be glad? The issue may sometimes be physically painful, but it is at all times metaphysically inspiring. 'Man is no more than the servant and interpreter of Nature,' said Sir Francis Bacon in his *Novum Organum*. And he added – 'Nature cannot be commanded except by being obeyed'. Such words are scarcely out of date.

BIBLIOGRAPHY

Abetti, G., *The Sun,* 1938, Lockwood.

Andrade, E. N. da C., *The Atom and its Energy,* 1947, Bell.
The New Chemistry, 1936, Bell.
Modern Physics, 1956, Bell.
Sir Isaac Newton, 1954, Collins.

Anthon's Classical Dictionary, Everyman.

Baker, Henry, *The Microscope Made Easy,* 1670.

Baker, Richard St Barbe, *Green Glory,* 1949, Wyn, N.Y.

Baring-Gould, S., *Curious Myths of the Middle Ages,* 1901, Longmans, Green.

Barlow, K. E., *The Discipline of Peace (2nd ed.),* 1971, Charles Knight.

Barret, W., and Besterman, T., *The Divining Rod,* 1926, Methuen.

Bates, H. W., *The Naturalist on the River Amazon,* 1930, E. P. Dutton, N.Y.

Beebe, W., *Adventuring with Beebe,* 1956, Bodley Head.
Half Mile Down, 1940, Bodley Head.

Bell, A. H., *Water Diviners and their Methods,* 1931, Bell.

Bentley, W. A., and Humphreys, W. J., *Snow Crystals,* 1931, McGraw-Hill.

Berry, E. W., *Tree Ancestors,* 1923, Williams & Wilkins, Baltimore.

Besterman, T., *Water-Divining,* 1938, Methuen.

Billings, H., *The Power and the Valley,* 1953, Hart-Davis.

Blake, E. H., *Drainage and Sanitation,* 1948, Batsford.

Bonner, J., and Galston, A. W. *Principles of Plant Physiology,* 1952, Freeman, San Francisco.

Bostrom, R. C., *Geology of Jan Mayen,* 1948, Penguin Books.

Botley, C. M., *The Air and its Mysteries,* 1938, Bell.

Botticher, C., *Der Baumkultus der Hellenen,* 1856, Weidmannsche Buchhandlung, Berlin.

Bowen, E. J., *Chemical Aspects of Light,* 1942, Clarendon Press.

Bower, F. O., *Botany of the Living Plant,* 1947, Macmillan.

Bragg, W., *Concerning the Nature of Things,* 1925, Bell.
The Universe of Light, 1947, Bell.

Breton, H. H., *The Great Blizzard of Christmas*, 1928, Hoyten & Cole, Plymouth.

Bridges, T. C., *Great Canals*, 1936, Nelson.

Broglie, Louis de, *The Revolution in Physics*, 1954, Routledge & Kegan Paul.

Brooks, C. F., *Why the Weather*, 1935, Chapman & Hall.

Browne, Sir Thomas, *Religio Medici*, 1643.

Bucke, R. M., *Cosmic Consciousness*, 1905, Philadelphia.

Cain, E. E., *Cyclone*, 1933, Stockwell.

Calder, Ritchie, *Man Against the Desert*, 1951, Allen & Unwin.

Campbell, D. H., *Structure and Development of Mosses and Ferns*, 1905, Macmillan.

Carpenter, E., *Pagan and Christian Creeds*, 1920, Allen & Unwin.

Casteret, N., *Ten Years Under the Earth*, 1939, Dent.
   My Caves, 1947, Dent.
   Cave Men, New and Old, 1951, Dent.

Catlin, G., *North American Indians*, 1913, Leary, Stuart & Co., Philadelphia.

Chase, Stuart, *Rich Land, Poor Land*, 1936, McGraw.

Cheskin, L., *Colours, What They Do For You*, 1947, Liveright, N.Y.

Chevalier, P., *Subterranean Climbers*, 1951, Dent.

Clements, J. B., *Water and the Land*, 1940, Oxford.

Cobbett, William, *Woodlands*, 1825.

Coleridge, S. T., *Nonesuch Edition*, edited by Stephen Potter.

Coles-Finch, W., *Water, Its Origin and Use*, 1908, Alston Rivers.
   Watermills and Windmills, 1933, Daniel.

Cook, W. A., *Through the Wilderness of Brazil by Horse, Canoe, and Float*, 1909, American Tract Society.

Coulter, John Merle, and Chamberlain, C. J., *The Morphology of Gymnosperms*, 1917, The University of Chicago Press.

Cox, George, *The Mythology of the Aryan Nations*, 1882.

Darwin, Charles, *Autobiography*, 1929, Watts.
   The Movement of Plants, 1880, Murray.
   The Origin of Species, 1951, Oxford.
   The Voyage of the Beagle, Everyman.

Davis, W. M., *Whirlwinds, Cyclones and Tornadoes*, 1884, Leed & Shepard, N.Y.

Dickinson, L., *The Greek Idea*, 1909, Methuen.

Dingle, Herbert, *Modern Astrophysics*, 1924, Collins.
   The Scientific Adventure, 1952, Putnam.

Dixon, H. H., *Transpiration and the Ascent of Sap*, 1910, Macmillan.

Douglas, A. C., *Gliding and Advanced Soaring*, 1947, Murray.

Duggar, Benjamin, *Plant Physiology*, 1911, Macmillan.

Eckerman. *Conversations with Goethe*, Everyman.

Eddington, A. S., *The Expanding Universe*, 1933, Cambridge University Press.

The Nature of the Physical World, 1928, Cambridge University Press.

The Philosophy of Physical Science, 1949, Cambridge University Press.

Ellis, Havelock, *The Dance of Life*, 1923, Constable.

Encyclopædia of Religion and Ethics, ed. James Hastings, 1908, Scribners.

Evans, Ifor, *Literature and Science*, 1954, Allen & Unwin.

Evelyn, John, *Sylva*, 1825, H. Colburn.

Fabre, J. H., *The Heavens*, 1924, Fisher Unwin.

The Wonder Book of Plant Life, 1924, Fisher Unwin.

This Earth of Ours, 1923, Fisher Unwin.

Farnell, L. R., *The Cults of the Greek States*, 1909, Clarendon Press.

Faulkner, E. H., *Ploughman's Folly*, 1944, Grosset, N.Y.

Finberg, J. G., *The Atom Story*, 1952, Wingate.

Fiske, J., *Myth and Myth Makers*, 1873, Boston.

Folkard, R., *Plant Lore, Legends and Lyrics*, 1884, Low, Marston, Searle & Rivington.

Forbes, J. D., *Observations on Glaciers*, 1842, Edinburgh.

Travels Through the Alps, 1900, Black.

Fox, Sir Cyril S., 'Finding Water', in *Science News 17*, 1950, Penguin Books.

Frazer, J. G., *The Golden Bough*, 1935, Macmillan.

The Dying God, 1936, Macmillan.

The Magic Art, 1936, Macmillan.

Balder the Beautiful, 1936, Macmillan.

Attis, Adonis, Osiris, 1936, Macmillan.

Free, E. E., and Hoke, Travis, *Weather*, 1929, Constable.

Freeman, A., *Goethe and Steiner*, 1947, Sheffield Educational Settlement.

Fritsch, F. E., and Salisbury, E., *Plant Form and Function*, 1946, Bell.

BIBLIOGRAPHY

Galton, F., *Narrative of an Explorer in Tropical South Africa,* 1853, Murray.

Gamow, George, *The Birth and Death of the Sun,* 1941, Macmillan.
Atomic Energy and Cosmic Human Life, 1947, Cambridge University Press.
Mr Tompkins Explores the Atom, 1945, Cambridge University Press.

Gardiner, Rolf, *England Herself; Ventures in Rural Restoration,* 1944, Faber & Faber.

Garnett, W., *A Little Book on Water Supply,* 1922, Cambridge.

Gassett, Ortega y, *The Revolt of the Masses,* 1951, Allen & Unwin.

Gatti, Attilio, *Great Mother Forest,* 1937, Scribner.

Geikie, Sir Archibald, *Text Book of Geology,* 1903, Macmillan.

Gili, Wyatt, *Myths and Songs from the South Pacific,* 1876, Beccles.

Goblet d'Alviella, Count E. F. A. *The Migration of Symbols,* 1894, Constable.

Goethe, W., *Theory of Colour,* 1840, Murray.

Grey, Sir George, *Polynesian Mythology and Ancient Traditional History of the New Zealanders,* 1906, E. P. Dutton.

Grimm, J., *Teutonic Mythology,* 1888, Bell.

Gubernatis, Angelo de, *La Mythologie des plantes,* 1882, Reinwald.

Gunther, K., and Deckert, K., *Creatures of the Deep Sea,* 1956, Allen & Unwin.

Hammer, C., *Goethe After Two Centuries,* 1952, State University Press, Louisiana.

Hardy, A., *The Open Sea: the World of Plankton,* 1956, Collins.

Harris, James Rendel, *The Ascent of Olympus,* 1917, Manchester University Press.

Hartley-Hennessy, T., *Healing by Water,* 1951, Daniel.

Hartridge, H., *Colours and How We See Them,* 1949, Bell.

Hartwig, G., *The Ariel World,* 1874, Longmans, Green.
The Subterranean World, 1871, Longmans, Green.

Harvey, E. N., *Living Light,* 1940, Princeton.

Haslett, A. W., 'Research Report', in *Science News 10,* 1949, Penguin Books.

Haviland, Maud D., *Forest, Steppe and Tundra,* 1926, Cambridge University Press.

Hecht, Selig, *Explaining the Atom,* 1955, Gollancz.

Heller, Erich, *The Disinherited Mind,* 1952, Bowes & Bowes.

Hendley, N. Ingram, 'Diatoms', in *Science News,* 6, 1948, Penguin Books.

BIBLIOGRAPHY

Herbert, A. S., *The Hot Springs of New Zealand*, 1921, Lewis.
Herodotus, Cary's Trans., *Book VI*, 1899, Appleton.
Hesiodus, *Hesiod, Works and Days*, 1932, Macmillan.
Homer, *The Odyssey*, 1948, Macmillan.
Hooke, Robert, *Microphalia*, 1660.
Hoyle, Fred., *The Nature of the Universe*, 1950, Blackwell.
Hughes-Gibb, E., *The Life Force in the Inorganic World*, 1930, Routledge & Kegan Paul.
Hull, Edward, *The Coalfields of Great Britain*, 1905, Rees.
Humboldt, A., *Examen Critique*, 1839, Gide.
  Aspects of Nature, 1849, Murray.
Humphrey, W., J., *Physics of the Air*, 1929, McGraw Hill.
Hurst, H. E., *The Nile*, 1952, Constable.
Huxley, Aldous, *Beyond the Mexique Bay*, 1934, Harper.
Huxley, T. H., 'On a Piece of Chalk', in *The Book of Naturalists*, 1944, Hale.
Jacks, G. V., *Soil*, 1954, Nelson.
Jacks, G. V., and Whyte, R. O., *Vanishing Lands*, 1939, Doubleday, Doran.
  The Raped Earth, 1939, Faber.
Jacks, L. P., *Near the Brink*, 1953, Allen & Unwin.
James, William, *Varieties of Religious Experience*, 1928, Longmans.
Jeans, James, *The Universe Around Us*, 1945, Cambridge University Press.
  Physics and Philosophy, 1948, Cambridge University Press.
Jeans, J. Stephen, *Waterways and Water Transport*, 1890, Spon.
Jeffries, Richard, *The Wood from the Trees*, 1945, Pilot Press.
Johnson, Amy, *Sunshine*, 1892.
Jones, E., *Essays in Applied Psychoanalysis*, 1951, Hogarth Press.
Jones, G. O., Rotblat, J., and Whitrow, C. J., *Atoms and the Universe*, 1956, Eyre and Spottiswoode.
Jung, C. G., *The Integration of the Personality*, 1949, Kegan, Paul.
Jungk, R., *Tomorrow is Already Here*, 1954, Hart-Davis.
Keary, C. F., *The Vikings in Western Christendom*, 1891, Putnam.
Keightley, Thomas, *The Mythology of Ancient Greece and Italy*, 1878, Appleton N.Y.
  The Fairy Mythology, 1900, Bell.
Keyserling, H., *South American Meditations on Hell and Heaven in the Soul of Man*, 1932, Harper.
Kimble, G., and Bush, R., *The Weather*, 1943, Penguin Books.

King, F. H., *The Soil*, 1899, Macmillan.
  Irrigation and Drainage, 1899, Macmillan.
Kingsley, M., *Travels in West Africa* (1879), reprint, 1972, Charles Knight.
Kraus, Eric, 'Physics of the Atmosphere' in *Science News 1*, 1946, Penguin Books.
  'Rain', in *Science News 5*, 1947, Penguin Books.
Lane, F. W., *The Elements Rage*, 1945, Country Life.
Lang, Andrew, *Myth Ritual and Religion*, 1899, Longmans, Green.
Lehns, Ernst, *Man or Matter*, 1951, Faber.
Lewes, G. H., *Life and Works of Goethe*, Everyman.
Lewis, Paul, *The Romance of Water Power*, 1931, Simpson, Low.
Lyell, Charles, *Principles of Geology*, 1840, Murray.
  A Manual of Elementary Geology, 1857, Appleton.
Luckiesh, M., *Artificial Light*, 1920, University of London Press.
Maby, J. C., and Franklin, T. B., *The Physics of the Divining Rod*, 1939, Bell.
MacDougal, D. T., *The Green Leaf*, New World of Science Series, 1930, Appleton-Century-Crofts.
Macmillan, H., *The First Forms of Vegetation*, 1874, Macmillan.
Mallory, W. H., *China, Land of Famine*, 1926, American Geographical Society.
Manly, G., *Climate and the British Scene*, 1952, Collins.
Mann, Ida, and Pirie, A., *The Science of Seeing*, 1946, Penguin Books.
Mannhardt, Wilhelm, *Wald and Feldkulte* (*Der Baumkultus der Germanen, etc.*), 1875, Gebruder Borntraeger.
Marsh, G. P., *Man and Nature* (Physical Geography as modified by human action), 1864, Sampson, Low.
Maspero, Sir Gaston C. C., *The Dawn of Civilisation*, 1922, Macmillan.
Massey, H. S. W., *Atoms and Energy*, 1953, Elek Books.
Massingham, Harold John, *The Tree of Life*, 1943, Chapman & Hall.
Maury, M. F., *Physical Geography*, 1908, American Book Co.
Maximov, N. A., *Plants in Relation to Water*, 1929, Allen & Unwin.
Maxwell, G., *In Malay Forests*, 1907, Blackwood.
McWilliams, Carey, *Ill Fares the Land*, 1942, Little, Brown, Boston.
Meyer, B. S., and Anderson, D. B., *Plant Physiology*, 1940, Chapman and Hall.

Miller, Hugh, *The Old Red Sandstone*, 1922, E. P. Dutton.

Milton, John, *Paradise Lost*.

Minnaert, M., *Light and Colour in the Open Air*, 1940, Bell.

Murray, Gilbert, *Four Stages of Greek Religion*, 1812, Columbia University Press.

Nature, Vol. 150, 1942.

New Schaff-Herzog Encyclopaedia of Religious Knowledge, ed. Sam Macaulay, Jackson, 1908, Funk & Wagnall.

Newton, Issac, *Optics*, Modern Edition, 1931, Bell.

Nicholson, H. A., *The Ancient Life History of the Earth*, 1898, Appleton.

Nicolson, Marjorie, *Newton Demands the Muse*, 1946, Princeton.
   The Microscope and English Imagination, College Studies in Modern Languages, Vol. 16, No. 4.
   The Telescope and Literature, Studies in Philology, 1935, Vol. XXXII.

O'Dea, John, *Darkness into Daylight*, Science Museum.
   The Social History of Lighting, 1958, Routledge & Kegan Paul.

O'Neill, John, *The Night of the Gods*, 1897, Quaritch.

Oakes, E. C., *Water Supplies through Three Centuries*, 1953, Preston.

Olcott, W. T., *Sun Lore in All Ages*, 1914, Putnam.

Olivier, C. P., *Meteors*, 1925, Williams and Wilkins, Baltimore.

Osborne, Fairfield, *Our Plundered Planet*, 1948, Little, Brown, Boston.
   The Limits of the Earth, 1954, Faber.

Ovid, *Metamorphoses*, 1946, Heinemann.

Pascal, Blaise, *Pensées*.

Paton, James, 'The Shower Cloud', in *Science News 17*, 1950, Penguin Books.

Pavan, Maria, 'Cave Science', in *Science News 5*, 1948, Penguin Books.

Peattie, Donald Culross, *The Road of a Naturalist*, 1941, Houghton Mifflin, Boston.
   Flowering Earth, 1948, Phoenix House.

Pentelow, F. T. K., *River Purification*, 1953, Arnold.

Person, H. S., *Little Waters*, 1936, U.S. Government Printing Office, Washington.

Perutz, D. M. F., 'Glaciers' in *Science News 6*, 1948, Penguin Books.

Pfeiffer, Ehrenfried, *The Earth's Face*, 1947, Faber.

Philpot, J. H., *The Sacred Tree,* 1897, Macmillan.

Phipson, T. L., *Phosphorescence,* 1870.

Physical Society, *Proceedings,* Vol. LV, 1943.

Pliny, *Natural History,* 1945, Harvard University Press.

Pope, Alexander, *Essay on Man,* 1733.

Porteous, A., *Forest Folklore, Mythology, and Romance,* 1928, Macmillan.

Radin, Paul, *The Story of the American Indian,* 1944, Liveright, N.Y.

Ralston, W. R. J., *Russian Folk Tales,* 1873, Smith, Elder.

Ratcliffe, Francis, N., *Flying Fox and Drifting Sand,* 1938, McBride, N.Y.

Raven-Hart, R., *Canoe Errant on the Nile,* 1936, Murray.

Reclus, E., *The Earth,* 1872, Harper.

Reid, C., *Submerged Forests,* 1913, Cambridge University Press.

Rinder, Frank, *Old World Japan,* 1896, Macmillan.

Rivera, J. E., *The Vortex (La Vorâgine),* 1935, Putnam.

Roberts, Michael, and Thomas, E. R., *Newton and the Origin of Colours,* 1934, Bell.

Robins, F. W., *The Story of Water,* 1946, Oxford.

Robins, John, *The Story of the Lamp.*

Rogers, F., and Beard, A., *Five Thousand Years of Glass,* 1948, Lippincott, N.Y.

Rolt, L. T. C., *Narrow Boat,* 1945, Eyre & Spottiswoode.
The Waterways of England, 1950, Allen & Unwin.

Rowe, W. H., *Our Forests,* 1947, Faber & Faber.

Ruskin, John, *Modern Painters,* Vol. IV, chaps. XII & XIII.
Fors Clavrigera, Letter XXXIV.
Arrows of the Chace, Letters on Science, 'Concerning Glaciers'.
Deucalion, Chap. III, 'Of Ice-cream'; Chap. IV, 'Labitur, et Labitur'; Chap. VI, 'Of Butter and Honey'.
The Eagle's Nest, 1870.
Ethics of the Dust, 1866.

Russell, Bertrand, *The ABC of Atoms,* 1924, Routledge & Kegan Paul.

Russell, Sir John, *Soil Condition and Plant Growth,* 1950, Longmans, Green.

Sachs, Julius von, *The Physiology of Plants,* 1887, Clarendon Press.

Salisbury, Rollin D., *Physiography,* 1907, Murray.

Saunders, B. C., and Clark, R. E. D., *Order and Chaos in the World of Atoms*, 1942, English Universities Press.

Schindler, Maria, *Goethe's Theory of Colour*, Steiner Book Centre.

Scoresby, W., *The Arctic Regions*, 1820.
Observations on a Greenland Voyage, 1810.

Scott, D. H., *Evolution of Plants*, 1911, Holt.

Sears, Paul, *Deserts on the March*, 1947, University of Oklahoma Press.

Seneca, Lucius Annaeus, *Letters to Lucilius*, 1932, Clarendon Press.

Seward, A. C., *Geology for Everyman*, 1944, Macmillan.

Shadwell, A., *The London Water Supply*, 1899, Longmans, Green.

Shelford, B., *Curiosities of Light*, 1899.

Sherrington, Charles, *Goethe on Nature and Science*, 1949, Cambridge University Press.

Skeat, W. W., *Malay Magic*, 1900, Macmillan.

Skrine, F. H. B., *The Heart of Asia*, 1899, Methuen.

Smith, B., *Geological Aspects of Underground Water Supplies*, 1935, Royal Society of Arts.

Smith, Robertson, *The Religion of the Semites*, 1926, Macmillan.
Lectures on the Religion of the Semites, 1927, Macmillan.

Spence-Hales and Bland, J., *England's Water Problem*, 1939, Country Life.

Spengler, O., *The Decline of the West*, 1945, Knopf.

Stanley, Sir Henry Morton, *In Darkest Africa*, 1913, Scribner's.

Steiger, G. N., *A History of the Far East*, 1944, Ginn & Co.

Steiner, Rudolf, *Colour*, Steiner Book Centre.

Stiles, W., *The Respiration of Plants*, 1952, Methuen.
Introduction to the Principles of Plant Physiology, 1950, Methuen,

Stopes, Marie C., *Ancient Plants*, 1910, Blackie.

Stopes, M. C., and Wheeler, R. V., *Monograph on the Constitution of Coal*, 1918, H.M.S.O.

Sullivan, J. W. N., *Isaac Newton*, 1938, Macmillan.
The Physical Nature of the Universe, 1932, Gollancz.
Science, a New Outline, 1935, Nelson.

Sykes, Friend, *Humus and the Farmer*, 1952, Rodale Press.

Tacitus, *Histories*, 1937, Putnam.

Talman, C., *The Realm of the Air*, 1931, Bobbs-Merrill.

Tannehill, I. R., *Drought*, Oxford.

Taylor, William Ling, *Forest and Forestry in Great Britain*, 1946, Lockwood.

BIBLIOGRAPHY

Tolman, C. F., *Ground Water*, 1937, McGraw-Hill.

Tomlinson, H. M., *The Sea and the Jungle*, 1920, Dutton.

Tompkins, B., *Springs of Water*, 1925, Hurst & Blackett.
   The Theory of Water Finding by the Divining Rod, 1899, Chippenham.

Timiriageff, C., 'The Cosmic Function of the Green Plant', *Proceedings*, Royal Society, 72, 424–61 : 1903.

Toynbee, A. J., *Study of History*, 1935, Oxford.

Trueman, A. E., *An Introduction to Geology*, 1938, Murby.

Turing, H. D., *River Pollution*, 1952, Arnold.
   Pollution, Four Reports prepared for the British Field Sports Society, 1947-9.

Tylor, Sir Edward B., *Primitive Culture*, 1924, Brentano.
   Researchers into the Early History of Mankind and the Development of Civilization, 1878, Estes and Lauriat, Boston.

Tyndall, John, *Forms of Water*, 1872, King & Co.
   Hours of Exercise in the Alps, 1899, Longmans, Green.
   The Glaciers of the Alps, 1860, Murray,
   Six Lectures on Light, 1895, Longmans.
   Notes on Light, 1869, Royal Institution

Tyrell, G. W., *The Earth and its Mysteries*, 1953, Bell.

Vinci, Leonardo da, *Notebooks*, 1938, Cape.

Virgil, *The Aeneid*, Heritage Press.

Vogt, William, *The Road to Survival*, 1948, William Sloane.

Wallis Budge, E. A., *Amulets and Superstitions*, 1930, Milford.

Ward, F. Kingdon, *About This Earth*, 1956, Cape.

Weeks, J. H., *Among Congo Cannibals*, 1913, Lippincott.

Weil, Simone, *The Need for Roots*, 1952, Routledge & Kegan Paul.

Wells, H. G., *The Outline of History*, 1921, Macmillan.

Wells, H. G., Huxley, Julian S., and Wells, G. P., *The Science of Life*, 1938, Doubleday Doran.

Whipple, G. C., *The Microscopy of Drinking Water*, 1907, Wiley, N. Y.
   Sewage Disposal, 1916, Wiley.
   The Value of Pure Water, 1927, Wiley.

Whitehead, A. N., *Science and the Modern World*, 1926, Cambridge University Press.

Wigglesworth, V. B. 'The Light of Glow-worms and Fire-flies', *Science News 11*, Penguin Books.

BIBLIOGRAPHY

Wilde, Lady Jane F., *Ancient Legends, Mystic Charms and Superstitions of Ireland,* 1925, Chatto & Windus.

Williams, S. Wells, *The Middle Kingdom,* 1883, W. H. Allen.

Wills, Philip, *On Being a Bird,* 1953, Parish.

Wilson, J. Leighton, *Western Africa,* 1856, Harper.

Wilson, M., *What is Colour?,* 1949, Goethean Science Foundation.

Wilson, R. A., *The Miraculous Birth of Language,* 1948, Philosophical Library, N.Y.

Zon, R., *Forests and Water in the Light of Scientific Investigation,* 1927, Government Printing Office, Washington.

INDEX

MORE ABOUT PENGUINS
AND PELICANS

Penguinews, which appears every month, contains details of all the new books issued by Penguins as they are published. From time to time it is supplemented by *Penguins in Print*, which is a complete list of all titles available. (There are some five thousand of these.)

A specimen copy of *Penguinews* will be sent to you free on request. For a year's issues (including the complete lists) please send 50p if you live in the British Isles, or 75p if you live elsewhere. Just write to Dept EP, Penguin Books Ltd, Harmondsworth, Middlesex, enclosing a cheque or postal order, and your name will be added to the mailing list.

In the U.S.A.: For a complete list of books available from Penguin in the United States write to Dept CS, Penguin Books Inc., 7110 Ambassador Road, Baltimore, Maryland 21207.

In Canada: For a complete list of books available from Penguin in Canada write to Penguin Books Canada Ltd, 41 Steelcase Road West, Markham, Ontario.

Laurie Lee

CIDER WITH ROSIE

Cider With Rosie puts on record the England we have traded for the petrol engine. Recalling life in a remote Cotswold village some fifty years ago, Laurie Lee conveys the semi-peasant spirit of a thousand-years-old tradition.

'This poet, whose prose is quick and bright as a snake . . . a gay, impatient, jaunty and in parts slightly mocking book; a prose poem that flashes and winks like a prism' – H. E. Bates in the *Sunday Times*

Not for sale in the U.S.A.

Flora Thompson

LARK RISE TO CANDLEFORD

A gipsy once told the fortune of Laura (as Flora Thompson called herself in the trilogy of books which appear together in this volume): 'You are going to be loved by people you've never seen and never will see.'

That prophecy came true when she published her endearing and precise record of country life at the end of the last century – a record in which she brilliantly engraves the fast-dissolving England of peasant, yeoman and craftsman and tints her picture with the cheerful courage and the rare pleasures that marked a self-sufficient world of work and poverty.

As H. J. Massingham has well said in his introduction: 'Flora Thompson possesses the attributes both of sympathetic presentation and literary power to such a degree of quality and beauty that her claims upon posterity can hardly be questioned.'

Ronald Blythe

AKENFIELD

'Here is a delectable book; a book to linger over and cherish, every page of which compels fresh thought' – *The Times*

Already a best-seller, Ronald Blythe's close-up of a Suffolk village has, for most readers, justified C. P. Snow's forecast that it would become a classic of its kind. Only a man born and bred in the county could, one feels, have extracted the confidences and revelations which fill these pages, as members of the community tell their personal stories.

The veteran of the First World War recounts his experiences in the trenches; the young farm-worker regrets the limited prospects on the land today; the retired district nurse, who started work in 1925 on £2 a week, recollects the local mistrust of a nurse in a car; the former army officer describes the village's reaction to his arrival and his determination to make good in farming.

And as they talk the whole village of Akenfield comes suddenly to life.

'One of the most absorbing books that I have read in the last ten years. A penetrating, extraordinarily unprejudiced, yet deeply caring account of modern rural life in England' – Angus Wilson.

For copyright reasons this edition is not for sale in the U.S.A.

John Stewart Collis

THE WORM FORGIVES THE PLOUGH

The Worm Forgives the Plough combines in one volume two previous books by John Stewart Collis, *While Following the Plough* and *Down to Earth*. The first is an account of this remarkable man's experience in agriculture when he worked on the land during the Second World War, the second of a series of meditations on such extraordinary things as the potato, the plough, the worm, the ant and the dunghill.

Collis is a unique synthesis of the scientist, the scholar, the practical man and the poet; and he has the two qualities found in the greatest writers on country matters: a deep love of his subject and a complete lack of sentimentality about it. *The Worm Forgives the Plough* is in the great tradition of *The Compleat Angler* and *Rural Rides*.

This volume is published simultaneously with *The Vision of Glory* by John Stewart Collis.

Not for sale in the U.S.A.